Executive Summary

By the third quarter of 2012, the United States had deployed more than 2.1 gigawatts (GWac[1]) of utility-scale solar generation capacity, with 4.6 GWac under construction as of August 2012 (SEIA 2012). Continued growth is anticipated owing to state renewable portfolio standards and decreasing system costs (DOE 2012a). One concern regarding large-scale deployment of solar energy is its potentially significant land use. Efforts have been made to understand solar land use estimates from the literature (Horner and Clark 2013); however, we were unable to find a comprehensive evaluation of solar land use requirements from the research literature. This report provides data and analysis of the land use associated with U.S. utility-scale[2] ground-mounted photovoltaic (PV) and concentrating solar power (CSP) facilities.

After discussing solar land-use metrics and our data-collection and analysis methods, we present total and direct land-use results for various solar technologies and system configurations, on both a capacity and an electricity-generation basis. The total area corresponds to all land enclosed by the site boundary. The direct area comprises land directly occupied by solar arrays, access roads, substations, service buildings, and other infrastructure. We quantify and summarize the area impacted, recognizing that the quality and duration of the impact must be evaluated on a case-by-case basis. As of the third quarter of 2012, the solar projects we analyze represent 72% of installed and under-construction utility-scale PV and CSP capacity in the United States. Table ES-1 summarizes our land-use results.

---

[1] All capacity-based land-use intensity figures in this study are expressed in terms of MWac or GWac. This is to maintain consistency within the paper because CSP power plants are rated in terms of MWac. The conversion factor between dc-rating and ac-rating is discussed in Section 3.
[2] We define utility-scale as greater than 1 MWdc for PV plants and greater than 1 MWac for CSP plants.

Table ES-1. Summary of Land-Use Requirements for PV and CSP Projects in the United States

Technology	Direct Area		Total Area	
	Capacity-weighted average land use (acres/MWac)	Generation-weighted average land use (acres/GWh/yr)	Capacity-weighted average land use (acres/MWac)	Generation-weighted average land use (acres/GWh/yr)
Small PV (>1 MW, <20 MW)	5.9	3.1	8.3	4.1
Fixed	5.5	3.2	7.6	4.4
1-axis	6.3	2.9	8.7	3.8
2-axis flat panel	9.4	4.1	13	5.5
2-axis CPV	6.9	2.3	9.1	3.1
Large PV (>20 MW)	7.2	3.1	7.9	3.4
Fixed	5.8	2.8	7.5	3.7
1-axis	9.0	3.5	8.3	3.3
2-axis CPV	6.1	2.0	8.1	2.8
CSP	7.7	2.7	10	3.5
Parabolic trough	6.2	2.5	9.5	3.9
Tower	8.9	2.8	10	3.2
Dish Stirling	2.8	1.5	10	5.3
Linear Fresnel	2.0	1.7	4.7	4.0

We found total land-use requirements for solar power plants to have a wide range across technologies. Generation-weighted averages for total area requirements range from about 3 acres/GWh/yr for CSP towers and CPV installations to 5.5 acres/GWh/yr for small 2-axis flat panel PV power plants. Across all solar technologies, the total area generation-weighted average is 3.5 acres/GWh/yr with 40% of power plants within 3 and 4 acres/GWh/yr. For direct-area requirements the generation-weighted average is 2.9 acres/GWh/yr, with 49% of power plants within 2.5 and 3.5 acres/GWh/yr. On a capacity basis, the total-area capacity-weighted average is 8.9 acres/MWac, with 22% of power plants within 8 and 10 acres/MWac. For direct land-use requirements, the capacity-weighted average is 7.3 acre/MWac, with 40% of power plants within 6 and 8 acres/MWac. Other published estimates of solar direct land use generally fall within these ranges.

Both capacity- and generation-based solar land-use requirements have wide and often skewed distributions that are not well captured when reporting average or median values. Some solar categories have relatively small samples sizes, and the highest-quality data are not available for all solar projects; both of these factors must be considered when interpreting the robustness of reported results. Owing to the rapid evolution of solar technologies, as well as land-use practices and regulations, the results reported here reflect past performance and not necessarily future trends. Future analyses could include evaluating the quality and duration of solar land-use impacts and using larger sample sizes and additional data elements to enable a thorough investigation of additional land-use factors.

Table of Contents

1 **Introduction** 1
2 **Solar Power Plant Land-Use Metrics** 2
3 **Solar Land-Use Data and Methodology** 4
4 **Results** 6
    4.1 Summary Results 7
    4.2 PV Land-Use Results 9
        4.2.1 Evaluation of PV Packing Factors 12
        4.2.2 Impact of Location and Tracking Configuration on PV Land Use 13
    4.3 CSP Land-Use Results 15
5 **Conclusions** 17
References 20
Appendix A. CSP Solar Multiple Ranges 22
Appendix B. PV Projects Evaluated 24
Appendix C. CSP Projects Evaluated 32
Appendix D. Impact of PV System Size and Module Efficiency on Land-Use Requirements 34
Appendix E. Impact of CSP System Size and Storage on Land-Use Requirements 37

List of Figures

Figure 1. NREL mesa top PV system—example of direct and total land use 3
Figure 2. Map of PV and CSP installations evaluated 7
Figure 3. Distribution of solar land-use requirements—whiskers indicate maximum and minimum values, box indicates 75th (top of box) and 25th (bottom of box) percentile estimates 8
Figure 4. Distribution of generation-based solar land-use requirements—whiskers indicate maximum and minimum values, box indicates 75th (top of box) and 25th (bottom of box) percentile estimates. Blue dot represents eSolar's Sierra Sun Tower (10 acres/GWh/yr), separated for clarity (but not considered an outlier) 9
Figure 5. Distribution of small PV land-use requirements—whiskers indicate maximum and minimum values, box indicates 75th (top of box) and 25th (bottom of box) percentile estimates 11
Figure 6. Distribution of large PV land-use requirements—whiskers indicate maximum and minimum values, box indicates 75th (top of box) and 25th (bottom of box) percentile estimates 12
Figure 7. Capacity-weighted average packing factor for PV projects evaluated—whiskers indicate maximum and minimum values, box indicates 75th (top of box) and 25th (bottom of box) percentile estimates 13
Figure 8. Modeled data showing relationship between CSP thermal storage and land-use intensity 16
Figure D-1. Total-area requirements for small PV installations as a function of PV plant size 34
Figure D-2. Total-area requirements for large PV installations as a function of PV plant size 35
Figure D-3. Capacity-based direct-area land-use requirements for all PV systems as a function of module efficiency 35
Figure D-4. Generation-based direct-area land-use requirements for all PV systems as a function of module efficiency 36
Figure E-1. Total-area requirements for CSP installations as a function of plant size 37
Figure E-2. Direct-area requirements for CSP installations as a function of plant size 38
Figure E-3. Total generation-based area requirements for CSP installations as a function of storage hours 38
Figure E-4. Total capacity-based area requirements for CSP installations as a function of storage hours. 39

List of Tables

Table ES-1. Summary of Land-Use Requirements for PV and CSP Projects in the United States v
Table 1. Summary of Data Categories Used for PV and CSP Plants 4
Table 2. Summary of Collected Solar Power Plant Data (as of August 2012) 6
Table 3. Total Land-Use Requirements by PV Tracking Type 10
Table 4. Direct Land-Use Requirements by PV Tracking Type 10
Table 5. Impacts of 1-Axis Tracking on Land-Use Intensity Compared With Fixed-Axis Mounting 14
Table 6. Total Land-Use Requirements by CSP Technology 15
Table 7. Direct Land-Use Requirements by CSP Technology 15
Table 8. Summary of Direct Land-Use Requirements for PV and CSP Projects in the United States 18
Table 9. Summary of Total Land-Use Requirements for PV and CSP Projects in the United States 19
Table A-1. CSP Solar Multiple Ranges and Corresponding Estimated Annual Generation Values 22
Table B-1. PV Land-Use Data 24
Table C-1. Concentrating Solar Power Land-Use Data 32

1 Introduction

By the third quarter of 2012, the United States had deployed more than 2.1 gigawatts (GWac[3]) of utility-scale solar generation capacity, with 4.6 GWac under construction as of August 2012 (SEIA 2012). Continued growth is anticipated owing to state renewable portfolio standards and decreasing system costs (DOE 2012a). One concern regarding large-scale deployment of solar energy is its potentially significant land use. Estimates of land use in the existing literature are often based on simplified assumptions, including power plant configurations that do not reflect actual development practices to date. Land-use descriptions for many projects are available from various permitting agencies and other public sources, but we were unable to locate a single source that compiles or summarizes these datasets. The existing data and analyses limit the effective quantification of land-use impacts for existing and future solar energy generation, particularly compared with other electricity-generation technologies.

This report provides data and analysis of the land use associated with U.S. utility-scale ground-mounted photovoltaic (PV) and concentrating solar power (CSP) facilities, defined as installations with capacities greater than 1 MW. The next section (Section 2) discusses standard land-use metrics and their applicability to solar power plants. We identify two major classes of solar plant land use—direct impact (disturbed land due to physical infrastructure development) and total area (all land enclosed by the site boundary)—by which we categorize subsequent results. Section 3 describes our solar land-use data collection and analysis methods. We derived datasets from project applications, environmental impact statements, and other sources and used them to analyze land use based on the capacity and generation of solar plants. Section 4 presents our results. In addition to summarizing PV and CSP land use, we examine relationships among land use, plant configuration, location, and technology. Finally, in Section 5, we identify limitations to the existing solar land-use datasets and suggest additional analyses that could aid in evaluating land use and impacts associated with the deployment of solar energy. Appendices include tables of our solar project data as well as more detailed analyses of specific land-use relationships.

---

[3] All capacity-based land-use intensity figures in this study are expressed in terms of MWac or GWac. This is to maintain consistency within the paper because CSP power plants are rated in terms of MWac. The conversion factor between dc-rating and ac-rating is discussed in Section 3.

2 Solar Power Plant Land-Use Metrics

There are many existing and proposed metrics for evaluating land-use impacts. Recent methods for quantifying land use include evaluating the direct and indirect life-cycle use (Fthenakis and Kim 2009) and assessing temporary and permanent land-area requirements (Denholm et al. 2009). While there is no single, generally accepted methodology (Canals et al. 2007), at least three general categories are used to evaluate land-use impacts: (1) the area impacted, (2) the duration of the impact, and (3) the quality of the impact (Koellner and Scholz 2008). The quality of the impact (also called the "damage function") evaluates the initial state of the land impacted and the final state across a variety of factors, including soil quality and overall ecosystem quality (Koellner and Scholz 2008).

This report closely follows the methodology outlined in a National Renewable Energy Laboratory (NREL) U.S. wind power land-use study (Denholm et al. 2009). We quantify and summarize the area impacted, recognizing that the quality and duration of the impact must be evaluated on a case-by-case basis. We consider two land-use metrics. The first is the total area, which corresponds to all land enclosed by the site boundary. The perimeter of this area is usually specified in blueprint drawings and typically fenced or protected. The second metric is the direct-impact area, which comprises land directly occupied by solar arrays, access roads, substations, service buildings, and other infrastructure. The direct-impact area is smaller than the total area and is contained within the total-area boundaries. Figure 1 illustrates the two types of areas, with the total area shaded yellow and the direct-impact area shaded orange.

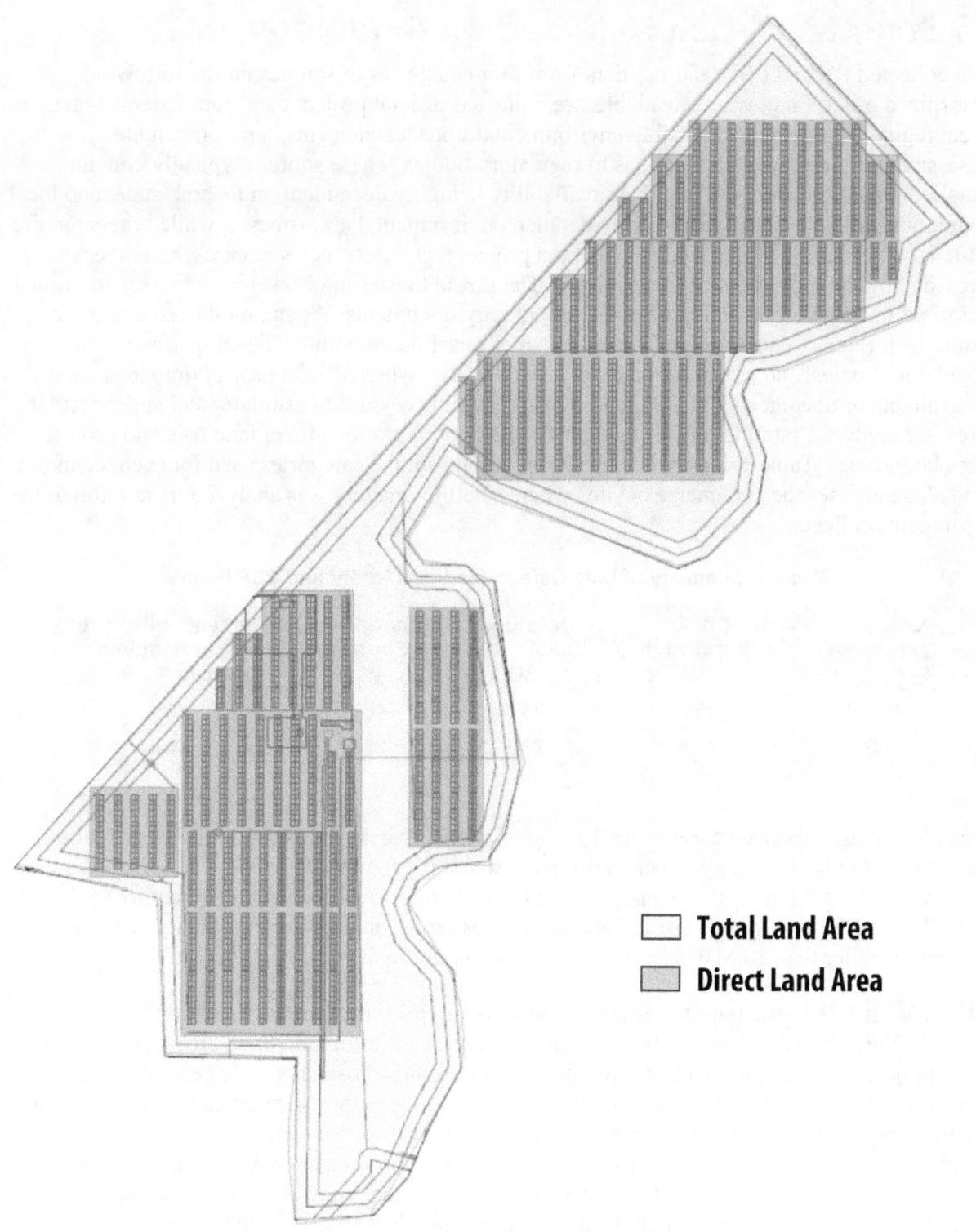

Figure 1. NREL mesa top PV system—example of direct and total land use[4]

---

[4] Access roads, infrastructure, and other direct impact areas are not shown in Figure 1.

3 Solar Land-Use Data and Methodology

We collected PV and CSP land-use data from four categories of sources, in the following prioritized order. First, where available, we collected official project data from federal, state, or local regulatory agencies, including environmental impact statements, environmental assessments, and project applications to regulatory bodies. These sources typically contain detailed project information, but their availability is highly dependent on federal, state, and local regulations as some states require very detailed environmental assessments, while others require little land-use analysis. Second, we collected project fact sheets, news releases, and other data provided by the project owner or developer. Data from these sources were used when additional information was needed and not found in regulatory documents. When no other source of data could be located, we used news articles, websites unaffiliated with the developer/owner or regulatory bodies, and other secondary sources. Finally, when official project drawings were unavailable or documents did not include information necessary to estimate total and direct land area, we analyzed satellite images to identify plant configuration, direct land use, and project-area boundaries. Table 1 shows the proportion of data source categories used for each technology and also indicates the percentage of sites where satellite imagery was analyzed in addition to the documents collected.

Table 1. Summary of Data Categories Used for PV and CSP Plants[5]

Technology	Official Documents (%)	Developer Documents (%)	Third-Party Sources (%)	Percent of Projects That Required Satellite Imagery
PV	18%	36%	46%	40%
CSP	44%	28%	28%	40%

For PV, we used these datasets to analyze the relationship between land-use intensity (defined as land use per unit of capacity or generation) and stated PV module efficiency, array configuration, and tracking type. For CSP, we analyzed the land-use intensity of several different technologies. For PV and CSP, we limited the analysis to systems larger than 1 MW in capacity. We classified systems smaller than 20 MW as "small" and those larger than 20 MW as "large."

We quantified land-use requirements on a capacity (area/MWac) and a generation (area/GWh/yr[6]) basis. Capacity-based results are useful for estimating land area and costs for new projects because power plants are often rated in terms of capacity. The generation basis provides a more consistent comparison between technologies that differ in capacity factor and enables evaluation of land-use impacts that vary by solar resource differences, tracking configurations, and technology and storage options. Most of the data collected for this analysis included the reported capacity of power plants but not annual generation. Because capacity-based land-use requirements are based on reported data, the capacity-based results are expected to have less uncertainty than the generation-based results.

---

[5] Percentages add up to over 100% because power plants evaluated with satellite imagery also required additional data sources to determine solar plant characteristics.

[6] Generation results are reported in area/(GWh per year) which we display as area/GWh/yr.

We simulated PV and CSP electricity generation using the System Advisor Model (SAM; Gilman and Dobos 2012). When available, we used project-specific inputs, such as location, array configuration, derate factor, and tracking technology. When project-specific inputs were unavailable, we used SAM default assumptions (e.g., if the tilt angle for fixed-tilt PV was unknown, we used SAM's latitude-tilt default assumption). The PV derate factor[7] was determined by dividing the AC reported capacity by the DC reported capacity for each project. The weighted-average derate factor (0.85) was used for projects that did not report both AC and DC capacity. All capacity-based land-use intensity figures in this study are expressed in terms of MWac. For CSP projects, a range of solar multiple[8] values was used to simulate annual generation output (see Appendix A for CSP solar multiple assumptions). Hourly solar resource and weather data for all projects were obtained from the NREL Solar Prospector tool[9] for each project's latitude and longitude. Each power plant was assigned to a cell within the National Solar Radiation Database (Wilcox 2007) equal in area to 0.1 degrees in latitude and longitude (approximately equal to a 10 km x 10 km square) (Perez et al. 2002). PV and CSP projects were simulated with typical direct-radiation-year weather data[10] (NREL 2012).

---

[7] The derate factor is used to determine the AC power rating at Standard Test Conditions (STC). The overall DC to AC derate factor accounts for losses from the DC nameplate power rating. We do not calculate the derate factor from component losses, but rather estimate the derate factor from the reported AC and DC power rating at each plant. For a discussion on derate factors, see
http://rredc.nrel.gov/solar/calculators/pvwatts/version1/change.html#derate (accessed April 2013).

[8] The solar multiple is the CSP field aperture area expressed as a multiple of the aperture area required to operate the power cycle at its design capacity (NREL 2012).

[9] The Solar Prospector is a mapping and analysis tool designed to provide access to geospatial data relevant to the solar industry. For more information, visit http://maps.nrel.gov/prospector (accessed May 2013).

[10] For consistency, PV and CSP data were both simulated using typical direct-radiation-year (TDY) weather data. Normally, CSP power plants are simulated using TDY data and PV power plants are simulated using typical meteorological year (TMY) data.

4 Results

We obtained land-use data for 166 projects completed or under construction (as of August 2012), representing 4.8 GWac of capacity, and 51 proposed projects, representing approximately 8 GWac of capacity (Table 2).

Table 2. Summary of Collected Solar Power Plant Data (as of August 2012)

	Small PV (<20 MW)		Large PV (>20 MW)		CSP	
	Projects	Capacity (MWac)	Projects	Capacity (MWac)	Projects	Capacity (MWac)
Completed	103	413	10	256	9	508
Under construction	17	165	20	1,846	7	1,610
Proposed	6	70	36	6,376	9	1,570
Total	**126**	**762**	**66**	**9,961**	**25**	**3,688**

We collected data on 4.8 GWac (72%) of the 6.7 GWac of completed or under-construction U.S. utility-scale solar capacity reported by SEIA (SEIA 2012). Figure 2 maps the solar projects evaluated. Appendix B and Appendix C detail all the projects and data sources. There are over 24 GWac of PV and CSP proposed (under development but not under construction) as of August 2012[11] (SEIA 2012), and the results reported in this study must be taken in light of a rapidly growing installed base. The results reported in this study reflect past performance and not necessarily future trends. For example, many of the largest PV systems currently proposed consist primarily of thin-film technology on fixed-tilt arrays, which may have different land use requirements than the results presented in this study.

---

[11] As of February 2013, there are 26 GWac of PV and CSP proposed (SEIA 2013).

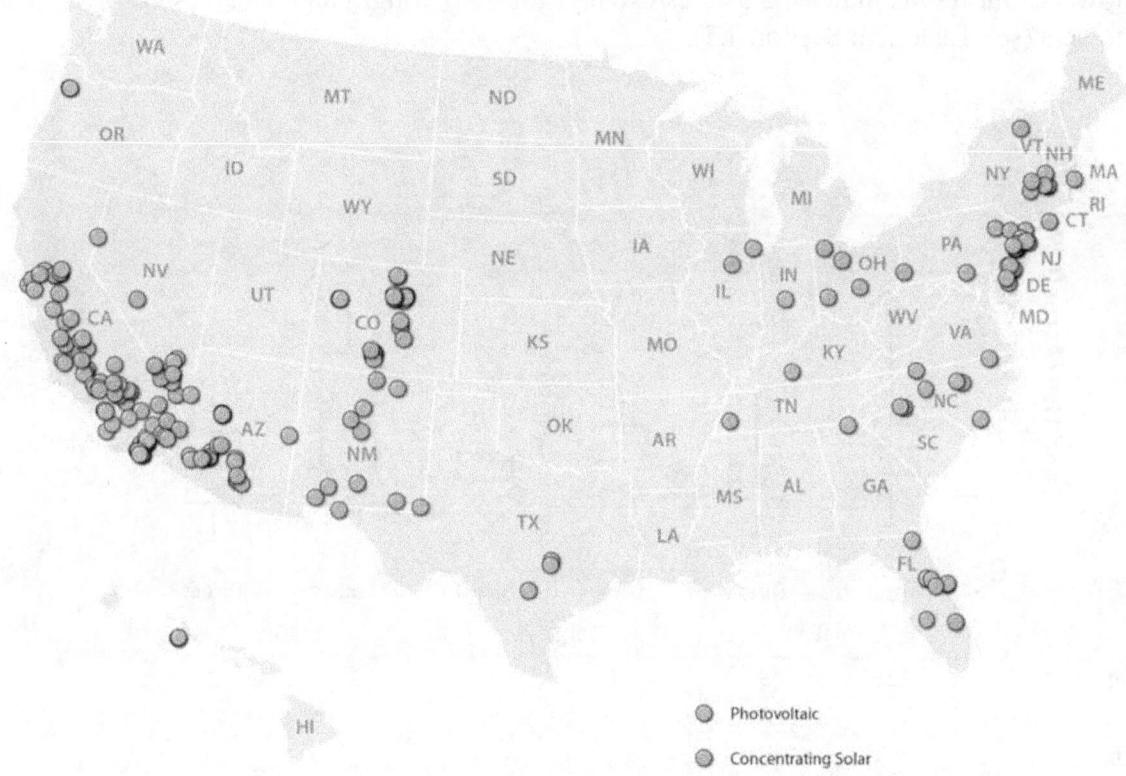

Figure 2. Map of PV and CSP installations evaluated

4.1 Summary Results

Figure 3 summarizes capacity-based total and direct land-use results for small and large utility-scale PV and CSP projects. Direct land-use requirements for small and large PV installations range from 2.2 to 12.2 acres/MWac, with a capacity-weighted average of 6.9 acres/MWac. Direct land-use intensity for CSP installations ranges from 2.0 to 13.9 acres/MWac, with a capacity-weighted average of 7.7 acres/MWac. Figure 4 shows generation-based total and direct land-use results. Direct land-use requirements for PV installations range from 1.6 to 5.8 acres/GWh/yr, with a generation-weighted average of 3.1 acres/GWh/yr. Direct land-use intensity for CSP installations ranges from 1.5 to 5.3 acres/GWh/yr, with a generation-weighted average of 2.7 acres/GWh/yr.

Solar direct land-use estimates in the literature generally fall within these ranges but are often smaller than the PV capacity-weighted averages we report and on par or larger for CSP capacity-weighted averages we report. Hand et al. (2012) estimate 4.9 acres/MWac for PV and 8.0 acres/MWac for CSP. Denholm and Margolis (2008) estimate 3.8 acres/MWac for fixed-tilt PV systems and 5.1 acres/MWac for 1-axis tracking PV systems. Our results indicate 5.5 acres/MWac for fixed-tilt PV and 6.3 acres/MWac for 1-axis tracking PV (capacity-weighted average direct land-use requirements for systems under 20 MW; see Table 4 in Section 4.2). Horner and Clark (2013) report 3.8 acres/GWh/yr for PV and 2.5 acres/GWh/yr for CSP. Fthenakis and Kim (2009) estimate 4.1 acres/GWh/yr for CSP troughs and 2.7 acres/GWh/yr for

CSP towers. Our results indicate 2.3 acres/GWh/yr for CSP troughs and 2.8 acres/GWh/yr for CSP towers (see Table 7 in Section 4.3).[12]

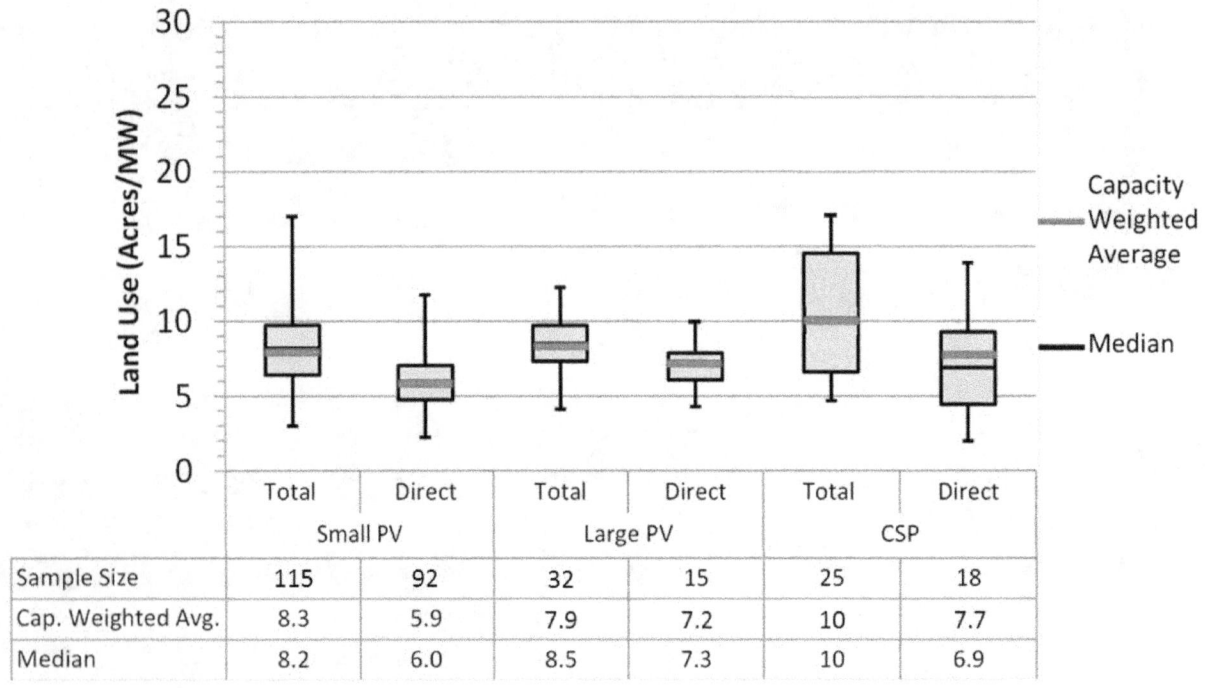

Figure 3. Distribution of solar land-use requirements—whiskers indicate maximum and minimum values, box indicates 75th (top of box) and 25th (bottom of box) percentile estimates

---

[12] Comparisons of generation-based land use results should be taken in light of the fact that annual generation (GWh) varies with solar resource (location). For example, generation-based results determined from solar power plants in a specific location may differ from results presented in this study, which includes solar plants from a variety of locations throughout the United States.

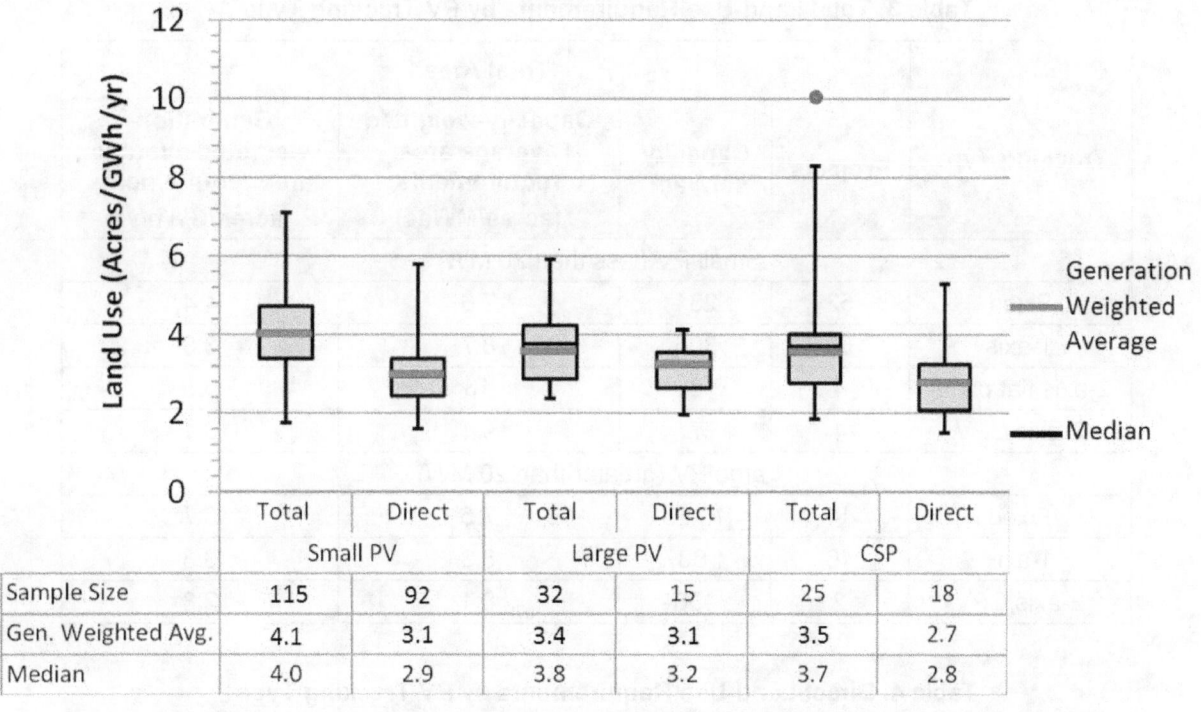

Figure 4. Distribution of generation-based solar land-use requirements—whiskers indicate maximum and minimum values, box indicates 75th (top of box) and 25th (bottom of box) percentile estimates. Blue dot represents eSolar's Sierra Sun Tower (10 acres/GWh/yr), separated for clarity (but not considered an outlier)

4.2 PV Land-Use Results

Table 3 and Table 4 summarize PV land requirements by tracking type for total and direct area, respectively. Total-area data were available for all systems evaluated; however, direct-area data were only available for a subset of these systems. Fixed-tilt and 1-axis PV systems account for a majority (96%) of projects evaluated.

On average, fixed-tilt systems use 13% less land than 1-axis tracking on a capacity basis but use 15% more land on a generation basis. This difference is due to increased generation resulting from tracking technologies. One-axis tracking systems can increase PV generation 12%–25% relative to fixed-tilt systems, and 2-axis tracking systems can increase PV generation by 30%–45% (Drury et al. 2012). We evaluated ten 2-axis PV plants: four flat panel (non-concentrating) projects and six concentrating PV (CPV) projects. Two-axis, flat panel systems appear to use more land than fixed and 1-axis plants on a capacity and generation basis, but general conclusions should not be drawn until the sample size is increased.

Table 3. Total Land-Use Requirements by PV Tracking Type[13]

Tracking Type	Total Area			
	Projects	Capacity (MWac)	Capacity-weighted average area requirements (acres/MWac)	Generation-weighted average area requirements (acres/GWh/yr)
Small PV (less than 20 MW)				
Fixed	52	231	7.6	4.4
1-axis	55	306	8.7	3.8
2-axis flat panel	4	5	13	5.5
2-axis CPV	4	7	9.1	3.1
Large PV (greater than 20 MW)				
Fixed	14	1,756	7.5	3.7
1-axis	16	1,637	8.3	3.3
2-axis CPV	2	158	8.1	2.8

Table 4. Direct Land-Use Requirements by PV Tracking Type[14]

Tracking Type	Direct Area			
	Projects	Capacity (MWac)	Capacity-weighted average area requirements (acres/MWac)	Generation-weighted average area requirements (acres/GWh/yr)
Small PV (less than 20 MW)				
Fixed	43	194	5.5	3.2
1-axis	41	168	6.3	2.9
2-axis flat panel	4	5	9.4	4.1
2-axis CPV	4	7	6.9	2.3
Large PV (greater than 20 MW)				
Fixed	7	744	5.8	2.8
1-axis	7	630	9.0	3.5
2-axis CPV	1	31	6.1	2.0

Figure 5 shows the capacity-based total and direct land-use requirement distributions for PV plants smaller than 20 MW. Direct land-use requirements for fixed-tilt PV installations range from 2.2 to 8.0 acres/MWac, with a capacity-weighted average of 5.5 acres/MWac. Direct land-use requirements for 1-axis tracking PV installations range from 4.2 to 10.6 acres/MWac, with a capacity-weighted average of 6.3 acres/MWac. Figure 6 shows the capacity-based total and

---

[13] Forty-two proposed projects representing 5,842 MWac could not be categorized by tracking type owing to insufficient information. These projects are not represented in this table.
[14] Forty-two proposed projects representing 5,842 MWac could not be categorized by tracking type due to insufficient information. These projects are not represented in this table.

direct land-use requirement distributions for PV plants larger than 20 MW. Relatively large deviations between the median and weighted average values are due to a few very large PV installations (over 100 MW) contributing heavily to weighted average results. We found that PV system size appears to have no significant impact on land-use requirements per unit of capacity (see Appendix D).

We also evaluated the impacts of efficiency on land-use intensity. We would expect land-use intensity to decrease with increasing module efficiencies, but we observed no significant trends between land-use intensity and module efficiency for small and large PV systems (see Appendix D). Variations in land-use intensity that remain after isolating for module efficiency and tracking type are not clearly understood. One source of variability could be the large range of packing factors described in the next section.

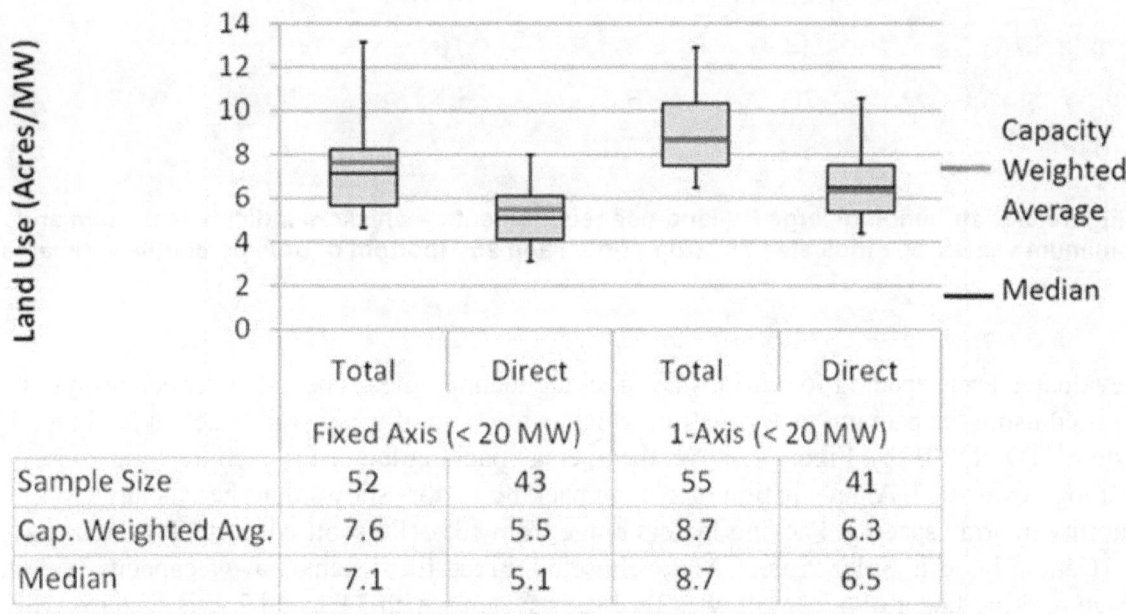

Figure 5. Distribution of small PV land-use requirements—whiskers indicate maximum and minimum values, box indicates 75th (top of box) and 25th (bottom of box) percentile estimates

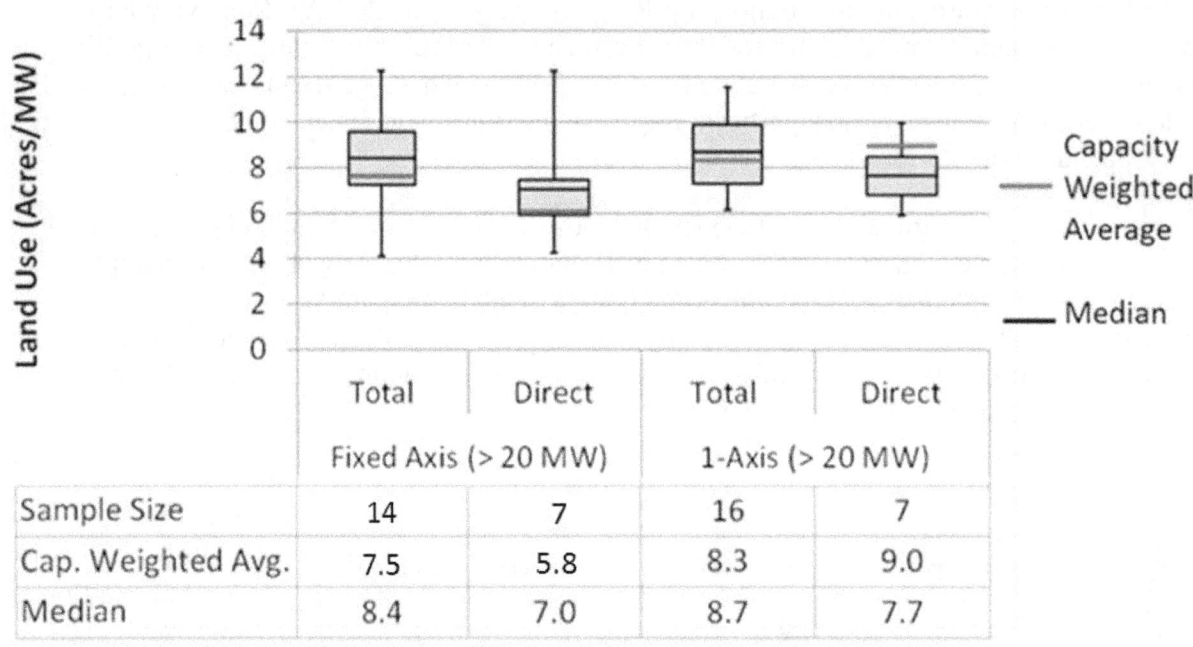

Figure 6. Distribution of large PV land-use requirements—whiskers indicate maximum and minimum values, box indicates 75th (top of box) and 25th (bottom of box) percentile estimates

4.2.1 Evaluation of PV Packing Factors

We evaluated array spacing for various PV tracking technologies. The area between arrays is quantified using the packing factor metric, which is the ratio of array area to actual land area for a system[15] (DOE 2012b). Figure 7 shows the average packing factor for each tracking technology evaluated. An evaluation of system packing factors shows that there is large variability in array spacing. Packing factors range from 13% (Prescott Airport CPV, Arizona) to 92% (Canton Landfill Solar Project, Massachusetts). Fixed-tilt systems have a capacity-weighted average packing factor of 47%, followed by 1-axis systems with 34% and 2-axis systems with 25%. Packing factor estimates from the research literature range from 20% to 67% (Horner and Clark 2013). The large variability in packing factor may contribute to the variability in land-use intensity observed, given an expectation that packing factor directly impacts land-use intensity. We did not attempt to isolate the impacts of packing factor, efficiency, capacity, and other factors on land-use intensity due to limited data availability. The availability of more data elements and larger sample sizes will enable a robust evaluation of these factors on land-use intensity.

---

[15] We display the packing factor ratio as a percentage. A 100% packing factor would represent complete coverage of solar panels with no spacing between arrays.

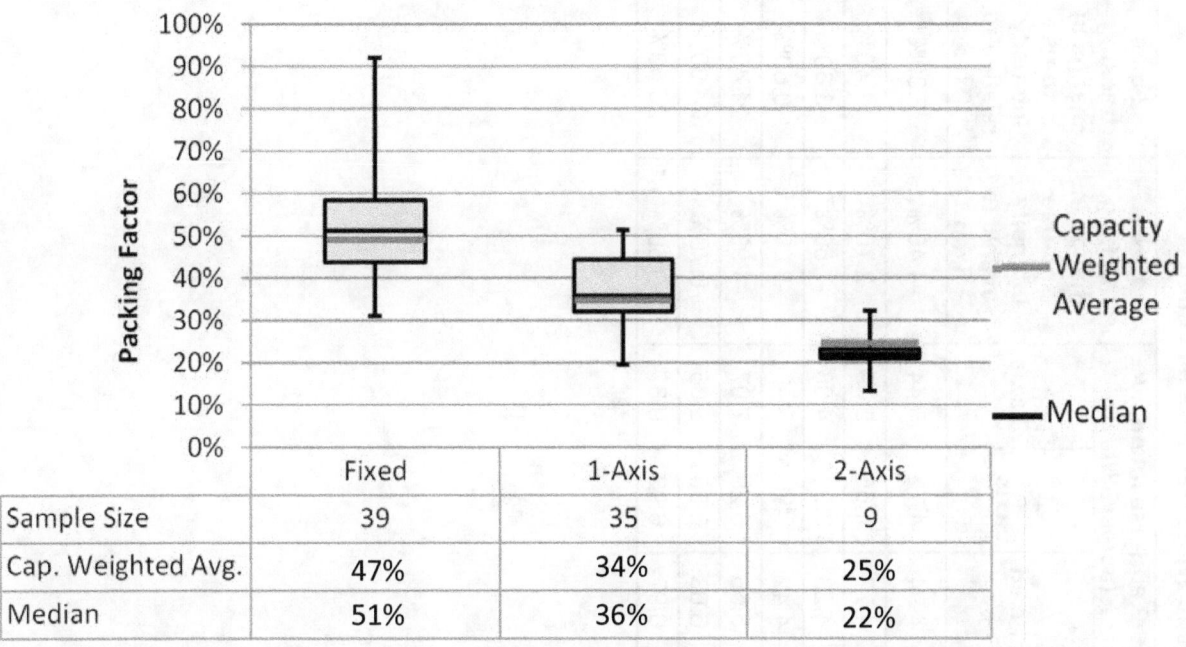

Figure 7. Capacity-weighted average packing factor for PV projects evaluated—whiskers indicate maximum and minimum values, box indicates 75th (top of box) and 25th (bottom of box) percentile estimates

4.2.2 Impact of Location and Tracking Configuration on PV Land Use

Given the relatively small amount of data, it is difficult to isolate the impact of any single factor on land-use requirements. This section isolates the theoretical impact of tracking arrays by simulating the performance of PV in multiple locations holding all other factors constant.

Table 5 summarizes the relative impacts of tracking on land-use intensity, simulated for a variety of locations throughout the United States. Although tracking systems generate more energy than fixed-tilt systems, they also require more land per unit of capacity, as shown in Section 4.2. We assume the capacity-weighted average land-use requirements (as reported in Table 4) for PV systems smaller than 20 MW when evaluating the impact of tracking arrays: 5.5 acres/MWac for fixed-tilt systems, 6.3 acres/MWac for 1-axis tracking systems, and 9.4 acres/MWac for 2-axis tracking systems. These results indicate that the expected increase in energy yield from 1-axis tracking systems (12%–22%) is partially countered by increases in land-use requirements per unit of capacity. While the land use per unit of generation generally decreases for 1-axis tracking systems compared with fixed-tilt systems, this metric generally increases for 2-axis tracking systems compared with fixed-tilt systems. This is because the spacing required for 2-axis tracking increases more than the relative increase in energy yield. The land-use advantage of 1-axis tracking is more pronounced in regions with higher direct normal irradiation (DNI) levels. Similarly, the negative land-use impacts of 2-axis tracking are less pronounced in regions with higher DNI levels. Denholm and Margolis (2008) estimated that land use per unit of generation would increase moving from fixed systems to 1-axis tracking systems and moving from fixed systems to 2-axis tracking systems.

Table 5. Impacts of 1-Axis Tracking on Land-Use Intensity Compared With Fixed-Axis Mounting

Region	Direct normal radiation (kWh/m2/yr)	Estimated energy production (kWh/kW)			1-axis tracking increase in energy yield relative to Fixed	2-axis tracking increase in energy yield relative to Fixed	Land-use intensity (acres/GWh/yr)			1-axis tracking change in land-use intensity relative to fixed	2-axis tracking change in land-use intensity relative to fixed
		Fixed	1-axis	2-axis			Fixed	1-axis	2-axis		
San Francisco, CA	1,883	1,551	1,828	1,951	17.9%	25.8%	4.94	4.72	5.44	-4.40%	9.30%
San Diego, CA	1,965	1,607	1,864	1,974	16.0%	22.8%	4.77	4.65	5.39	-2.70%	11.40%
Alamosa, CO	2,530	1,813	2,200	2,606	21.3%	43.7%	4.23	3.93	4.08	-7.50%	-3.60%
Phoenix, AZ	2,519	1,733	2,113	2,419	21.9%	39.6%	4.42	4.1	4.4	-8.00%	-0.60%
Jacksonville, FL	1,507	1,380	1,634	1,504	18.4%	9.0%	5.56	5.29	7.07	-4.90%	21.40%
Newark, NJ	1,263	1,268	1,422	1,321	12.1%	4.2%	6.03	6.08	8.06	0.70%	24.90%
Seattle, WA	1,112	1,100	1,249	1,136	13.5%	3.3%	6.97	6.92	9.37	-0.60%	25.50%

4.3 CSP Land-Use Results

Table 6 and Table 7 summarize total and direct land-use requirements by CSP technology, respectively. Note there are significantly fewer CSP projects in the United States than PV projects, and due to reliance on solar DNI resource, most CSP projects are in the Southwest (Figure 2). We collected data for 25 CSP projects, with only one linear Fresnel project and one dish Stirling project. It is more important to evaluate CSP in terms of land use per unit of generation because of the effect of storage and solar multiple, which can increase the amount of energy produced per unit of capacity (Turchi et al. 2010). Direct land-use requirements for CSP trough technology range from 2.0 to 4.5 acres/GWh/yr, with a generation-weighted average of 2.5 acres/GWh/yr. Direct land-use requirements for CSP tower technology range from 2.1 to 5.3 acres/GWh/yr, with a generation-weighted average of 2.8 acres/GWh/yr. We found that system size appears to have little impact on generation-based CSP land-use requirements (see Appendix E).

Table 6. Total Land-Use Requirements by CSP Technology

Technology	Total Area			
	Projects	Capacity (MWac)	Capacity-weighted average area requirements (acres/MWac)	Generation-weighted average area requirements (acres/GWh/yr)
All	25	3,747	10	3.5
Trough	8	1,380	9.5	3.9
Tower	14	2,358	10	3.2
Dish Stirling	1	2	10	5.3
Linear Fresnel	1	8	4.7	4.0

Table 7. Direct Land-Use Requirements by CSP Technology

Technology	Direct Area			
	Projects	Capacity (MWac)	Capacity-weighted average area requirements (acres/MWac)	Generation-weighted average area requirements (acres/GWh/yr)
All	18	2,218	7.7	2.7
Trough	7	851	6.2	2.5
Tower	9	1,358	8.9	2.8
Dish Stirling	1	2	2.8	1.5
Linear Fresnel	1	8	2.0	1.7

Data for CSP with multi-hour energy storage were also collected. Eight facilities included thermal storage technology, ranging from 3 to 15 hours of storage. One of the eight CSP facilities with storage is a parabolic trough system, while the remaining seven are tower systems. Little correlation is observed between storage and land-use intensity, both on a capacity and generation basis (see Appendix E). We would expect to see a trend of decreasing generation-based land use with increasing storage and increasing capacity-based land use with increasing storage based on modeled results as shown in Figure 8 (Turchi et al. 2010). Given the relatively small amount of data, it is difficult to isolate the impact of any single factor on land-use requirements. Higher sample sizes and additional data elements will enable a more robust evaluation of CSP land use.

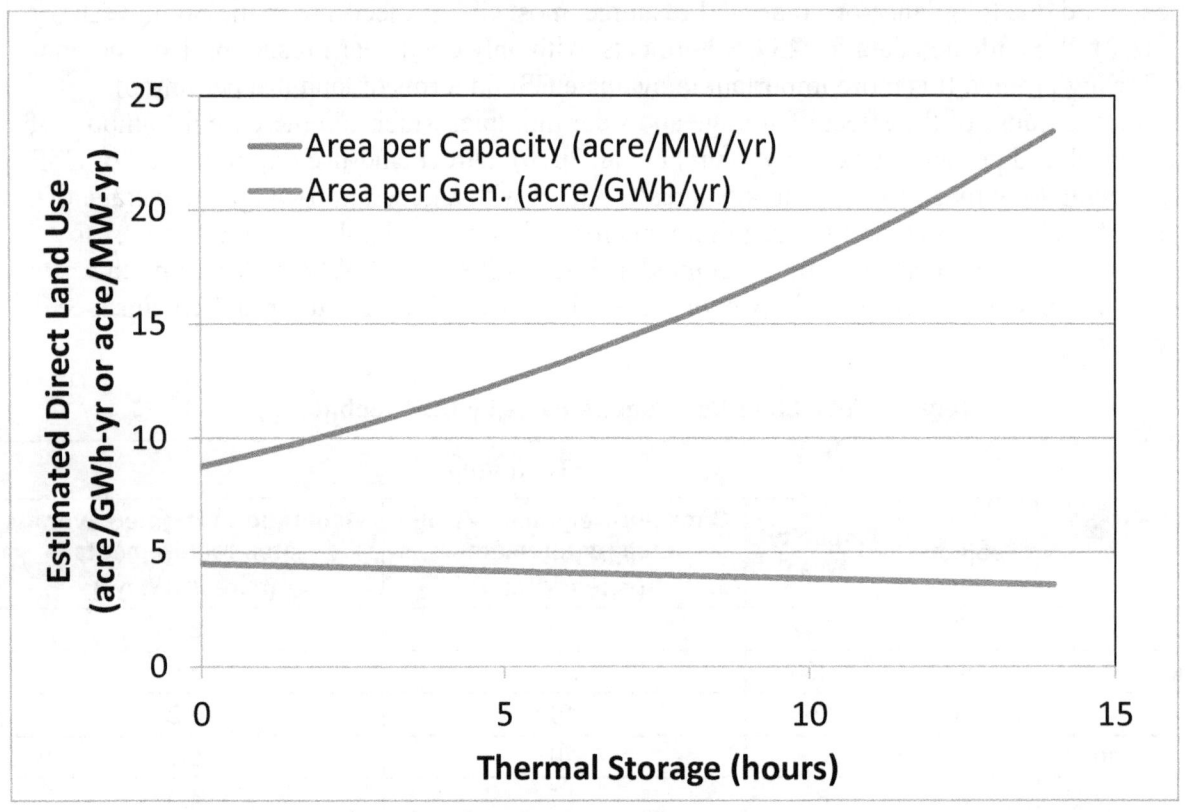

Figure 8. Modeled data showing relationship between CSP thermal storage and land-use intensity

Source: Turchi et al. 2010

5 Conclusions

Table 8 and Table 9 summarize the U.S. utility-scale PV and CSP land-use requirements evaluated in this report. Average total land-use requirements are 3.6 acres/GWh/yr for PV and 3.5 acres/GWh/yr for CSP. Average direct-area requirements are 3.1 acres/GWh/yr for PV and 2.7 acres/GWh/yr for CSP. On a capacity basis, the total-area capacity-weighted average for all solar power plants is 8.9 acres/MWac, with 22% of plants within 8 and 10 acres/MWac. For direct land-use requirements, the capacity-weighted average is 7.3 acre/MWac, with 40% of power plants within 6 and 8 acres/MWac. Solar land-use estimates from the literature generally fall within these ranges. Within the broad technology categories of PV and CSP, land-use metrics are also impacted by specific technology choices, such as cell efficiency, tracking method, and inclusion of thermal energy storage, and are a function of the solar resource available at each site.

Although our results stem from an empirically based effort to estimate solar land use, several caveats are warranted. Some solar-technology categories have relatively small samples sizes, which must be considered when interpreting the robustness of reported results. Over 26 GWac of PV and CSP are under development as of February 2013 (SEIA 2013), and the results reported in this study must be understood in light of a rapidly growing installed base. Additionally, various data sources were used when gathering information about solar projects. Although we tried to obtain the highest-quality sources (project applications and regulatory documents, referred to as "official documents" in this report), we collected official documents for only 20% of all projects evaluated. Other data sources are expected to have higher levels of uncertainty (although how much higher is unclear), which could contribute to the observed variability in results. With the exception of a few CSP projects, we collected reported capacity of power plants but not annual generation. The generation-based land-use results are expected to have higher levels of uncertainty because annual generation is simulated. Although generation-based results provide a more consistent approach when comparing land-use requirements across technologies, capacity-based results are useful for estimating land area and costs for new projects because power plants are often rated in terms of capacity. Finally, owing to the rapid evolution of solar technologies as well as land-use practices and regulations, the results reported here reflect past performance and not necessarily future trends.

We analyze elements that affect the area of solar impact, but we recognize that the duration of use and impact on land quality are also important when considering land use impacts. Future analyses could include evaluating the quality of land impacts, assessing both the initial state of the land impacted and the final states across a variety of factors, including soil quality and overall ecosystem quality. Finally, larger sample sizes and additional data elements would improve the robustness of the conclusions and enable a more thorough investigation of the impacts of additional factors, such as tilt angle, azimuth, PV module technology, CSP solar multiple, and storage technologies.

Table 8. Summary of Direct Land-Use Requirements for PV and CSP Projects in the United States

Technology	Number of projects analyzed	Capacity for analyzed projects (MWac)	Direct Area			
			Capacity-weighted average land use (acres/MWac)	Capacity-weighted average land use (MWac/km²)	Generation-weighted average land use (acres/GWh/yr)	Generation-weighted average land use (GWh/yr/km²)
Small PV (>1 MW, <20 MW)	92	374	5.9	42	3.1	81
Fixed	43	194	5.5	45	3.2	76
1-axis	41	168	6.3	39	2.9	86
2-axis flat panel	4	5	9.4	26	4.1	60
2-axis CPV	4	7	6.9	36	2.3	105
Large PV (>20 MW)	15	1,405	7.2	34	3.1	80
Fixed	7	744	5.8	43	2.8	88
1-axis	7	630	9.0	28	3.5	71
2-axis CPV	1	31	6.1	41	2.0	126
CSP	18	2,218	7.7	32	2.7	92
Parabolic trough	7	851	6.2	40	2.5	97
Tower	9	1,358	8.9	28	2.8	87
Dish Stirling	1	2	2.8	88	1.5	164
Linear Fresnel	1	8	2.0	124	1.7	145

Table 9. Summary of Total Land-Use Requirements for PV and CSP Projects in the United States

Technology		Number of projects analyzed	Capacity for analyzed projects (MWac)	Total Area			
				Capacity-weighted average land use (acres/MWac)	Capacity-weighted average land use (MWac/km^2)	Generation-weighted average land use (acres/GWh/yr)	Generation-weighted average land use (GWh/yr/km^2)
Small PV (>1 MW, <20 MW)		115	550	8.3	30	4.1	61
	Fixed	52	231	7.6	32	4.4	56
	1-axis	55	306	8.7	29	3.8	66
	2-axis flat panel	4	5	13	19	5.5	45
	2-axis CPV	4	7	9.1	27	3.1	80
Large PV (>20 MW)		32	3,551	7.9	31	3.4	72
	Fixed	14	1,756	7.5	33	3.7	67
	1-axis	16	1,637	8.3	30	3.3	76
	2-axis CPV	2	158	8.1	31	2.8	89
CSP		25	3,747	10	25	3.5	71
	Parabolic trough	8	1,380	9.5	26	3.9	63
	Tower	14	2,358	10	24	3.2	77
	Dish Stirling	1	2	10	25	5.3	46
	Linear Fresnel	1	8	4.7	53	4.0	62

References

Canals, L.M.; Bauer, C.; Depestele, J.; Dubreuil, A.; Freiermuth Knuchel, R.; Gaillard, G.; Michelsen, O.; Müller-Wenk. R.; Rydgren, B. (2007). "Key Elements in a Framework for Land Use Impact Assessment in LCA." *The International Journal of Life Cycle Assessment* (12:1); pp. 5–15.

Denholm, P.; Hand, M.; Jackson, M.; Ong, S. (2009). "Land-Use Requirements of Modern Wind Power Plants in the United States." NREL/TP-6A2-45834. Golden, CO: National Renewable Energy Laboratory.

Denholm, P.; Margolis, R. (2008). "Land-Use Requirements and the Per-Capita Solar Footprint for Photovoltaic Generation in the United States." *Energy Policy* (36:9); pp. 3531–3543.

DOE (U.S. Department of Energy). (2012a). *SunShot Vision Study*. DOE/GO-102012-3037. Accessed July 2012: http://www1.eere.energy.gov/solar/pdfs/47927.pdf.

DOE. (2012b). Solar Energy Glossary. Accessed August 2012: http://www1.eere.energy.gov/solar/sunshot/glossary.html.

Drury, E.; Lopez, A.; Denholm, P.; Margolis, R. (2012). "Relative Performance of Tracking versus Fixed Tilt Photovoltaic Systems in the United States." Golden, CO: National Renewable Energy Laboratory.

Fthenakis, V.; Kim, H.C. (2009). "Land Use and Electricity Generation: A Life-Cycle Analysis." *Renewable and Sustainable Energy Reviews* (13); pp. 1465–1474.

Gilman, P.; Dobos, A. (2012). *System Advisor Model, SAM 2011.12.2: General Description*. NREL/TP-6A20-53437. Golden, CO: National Renewable Energy Laboratory.

Hand, M.M.; Baldwin, S.; DeMeo, E.; Reilly, J.M.; Mai, T.; Arent, D.; Porro, G.; Meshek, M.; Sandor, D. (2012). *Renewable Electricity Futures Study*. eds. 4 vols. NREL/TP-6A20-52409. Golden, CO: National Renewable Energy Laboratory.

Horner, R.; Clark, C. (2013). "Characterizing variability and reducing uncertainty in estimates of solar land use energy intensity." *Renewable and Sustainable Energy Reviews* (23); pp. 129–137.

Koellner, T.; Scholz, R. (2008). "Assessment of Land Use Impacts on the Natural Environment." *The International Journal of Life Cycle Assessment* (13:1); pp. 32–48.

NREL (National Renewable Energy Laboratory). (2012). System Advisor Model User Documentation. https://www.nrel.gov/analysis/sam/help/html-php/.

Perez, R.; Ineichen, P.; Moore, K.; Kmiecikm, M.; Chain, C.; George, R.; Vignola, F. (2002). "A New Operational Satellite-to-Irradiance Model." *Solar Energy* (73:5); pp. 307–317.

SEIA (Solar Energy Industries Association). (2012). *Utility-Scale Solar Projects in the United States: Operating, Under Construction, or Under Development (Updated August 15, 2012)*. Washington, DC: SEIA.

SEIA. (2013). *Utility-Scale Solar Projects in the United States: Operating, Under Construction, or Under Development (Updated February 11, 2013)*. Washington, DC: SEIA.

Turchi, C.; Mehos, M.; Ho, C.; Kolb, G. (2010). "Current and Future Costs for Parabolic Trough and Power Tower Systems in the US Market." NREL/CP-5500-49303. Golden, CO: National Renewable Energy Laboratory.

Wilcox, S. (2007). *National Solar Radiation Database 1991 – 2005 Update: User's Manual*. NREL/TP-581-41364. Golden, CO: National Renewable Energy Laboratory.

Appendix A. CSP Solar Multiple Ranges

For CSP projects, a range of solar multiple values were used to simulate annual generation output. Assumptions for CSP solar multiple ranges are shown in Table A-1.

Table A-1. CSP Solar Multiple Ranges and Corresponding Estimated Annual Generation Values

Name	State	Storage (hours)	Solar multiple low	Solar multiple high	Estimated generation low (GWh/yr)	Estimated generation high (GWh/yr)
Crossroad Solar	AZ	10	2.2	2.8	683	822
Quartzsite	AZ	10	2.2	2.8	489	578
Saguaro Power Plant	AZ	0	1.1	1.4	2	2
Solana	AZ	6	1.9	2.4	992	1,155
Abengoa Mojave	CA	0	1.1	1.4	520	645
Coalinga	CA	0	1.1	1.4	9	28
Ford Dry Lake (Genesis)	CA	0	1.1	1.4	480	617
Hidden Hills 1	CA	0	1.1	1.4	545	655
Hidden Hills 2	CA	0	1.1	1.4	545	655
Ivanpah (all)	CA	0	1.1	1.4	869	1,024
Kimberlina	CA	0	1.1	1.4	9	11
Palmdale Hybrid Plant	CA	0	1.1	1.4	107	138
Rice Solar	CA	7	1.8	2.2	541	692
Rio Mesa 1	CA	0	1.1	1.4	529	659
Rio Mesa 2	CA	0	1.1	1.4	529	659
Rio Mesa 3	CA	0	1.1	1.4	529	659
SEGS (all)	CA	0	1.1	1.4	725	888
Solar Two	CA	3	1.3	1.7	20	30
Victorville 2 hybrid	CA	0	1.1	1.4	101	125
Saguache Solar	CO	15	2.6	3.2	1,073	1,216
Martin Next Generation	FL	0	1.1	1.4	71	105

Name	State	Storage (hours)	Solar multiple low	Solar multiple high	Estimated generation low (GWh/yr)	Estimated generation high (GWh/yr)
Nevada Solar One	NV	0.5	1.2	1.5	114	144
Tonopah (Crescent Dunes)	NV	10	2.2	2.8	525	590
Crossroad Solar	AZ	10	2.2	2.8	683	822
Quartzsite	AZ	10	2.2	2.8	489	578

Appendix B. PV Projects Evaluated

Table B-1. PV Land-Use Data

Asterisks represent data calculated from power plants that reported only AC capacity, as described in Section 3.

Name	State	MW - DC	Total area (acres)	Direct area (acres)	Tracking	Module efficiency	Status as of August 2012	Data source
Prescott Airport (CPV)	AZ	0.2	1.9	1.0	2 axis CPV	29%	Complete	Third party
Pima County Wastewater	AZ	1.1	8.4	6.4	1 axis	14%	Complete	Developer
Johnson Utilities	AZ	1.1	10.6	7.5	1 axis		Complete	Third party
Prescott Airport (1-Axis Phase 1)	AZ	2.8	22.6	22.3	1 axis	11%	Complete	Third party
Springerville	AZ	6.5	85.2	45.3	fixed	11%	Complete	Developer
Kingman Plant	AZ	10.0	70.5		1 axis	14%	Construction	Third party
Prescott Airport (1-Axis Phase 2)	AZ	11.8	94.0		1 axis		Construction	Third party
Luke Air Force Base	AZ	15.0	107.1		1 axis	19%	Complete	Third party
Hyder Plant	AZ	17.0	152.7		1 axis	14%	Construction	Third party
Paloma Plant	AZ	20.3*	234.9		fixed	11%	Complete	Third party
Cotton Center Plant	AZ	21.0	169.2		1 axis		Complete	Third party
Copper Crossing Solar Ranch	AZ	23.5*	169.1	139.1	1 axis	19%	Complete	Developer
Chino Plant	AZ	23.5*	187.9	164.4	1 axis	14%	Construction	Official documents
Tucson Solar	AZ	25.0	233.7		1 axis		Construction	Developer
Avra Valley	AZ	30.5*	352.4		1 axis	11%	Construction	Developer
Mesquite Solar 1	AZ	170.0	1,020.0		Unknown	15%	Construction	Official documents
Agua Caliente	AZ	340.6*	2,818.9		fixed	11%	Construction	Developer
Sonoran Solar Energy Project	AZ	352.4*	2,364.3		1 axis		Proposed	Official documents
Mesquite Solar Total	AZ	700.0	4,698.1		Unknown		Proposed	Third party
Western Riverside County Regional Wastewater Authority	CA	1.0	11.2	10.6	1 axis	20%	Complete	Developer
The North Face PV Plant	CA	1.0	5.9	5.9	1 axis		Complete	Third party

Name	State	MW - DC	Total area (acres)	Direct area (acres)	Tracking	Module efficiency	Status as of August 2012	Data source
Inlands Empire Utility Solar Farm	CA	1.0	12.6	8.9	1 axis	20%	Complete	Developer
West County Waste Water PV Plant	CA	1.0	11.7	6.9	2 axis flat	14%	Complete	Developer
Nichols Farms PV Plant	CA	1.0	8.0	8.0	2 axis CPV	25%	Complete	Third party
Budweiser PV Plant	CA	1.1	9.4	7.2	1 axis	15%	Complete	Official documents
Wal-Mart Apple Valley PV Plant	CA	1.1	10.7	7.8	1 axis	15%	Complete	Official documents
Rancho California PV Plant	CA	1.1	13.6	8.9	1 axis	19%	Complete	Developer
Hayward Wastewater PV Plant	CA	1.2	13.2	8.6	1 axis	14%	Complete	Third party
Chuckawalla State Prison PV Plant	CA	1.2	8.4	4.8	fixed	14%	Complete	Official documents
Ironwood State Prison PV Plant	CA	1.2	14.4	9.0	1 axis	13%	Complete	Official documents
Sacramento Soleil	CA	1.3	10.0	8.1	fixed	11%	Complete	Developer
USMC 29 Palms	CA	1.3	10.6	7.0	fixed		Complete	Developer
Box Canyon Camp Pendleton	CA	1.4	9.6	5.6	fixed	14%	Complete	Third party
Vaca-Dixon Solar Station	CA	2.6	17.8	11.5	fixed	14%	Complete	Developer
Newberry Springs PV Plant	CA	3.0	25.8		1 axis		Proposed	Third party
Sunset Reservoir	CA	5.0	15.3	15.3	fixed		Complete	Third party
Aero Jet Solar Project	CA	6.0	47.0	32.3	1 axis		Complete	Developer
CALRENEW-1	CA	6.2	60.4	46.5	fixed	9%	Complete	Third party
Porterville Solar Plant	CA	6.8	37.6	31.4	fixed	14%	Complete	Third party
Palm Springs project 1	CA	8.0	42.9		1 axis	14%	Construction	Third party
Dillard Solar Farm	CA	12.0	94.3	70.4	1 axis	15%	Complete	Developer
China Lake PV Plant	CA	13.8	138.6		1 axis	20%	Construction	Third party
Bruceville Solar Farm	CA	16.4	131.1	92.9	1 axis	15%	Complete	Official documents
Kammerer Solar Farm	CA	16.6	129.1	111.1	1 axis	15%	Complete	Official documents
Antelope Solar Farm	CA	20.0	234.9		Unknown		Proposed	Developer
Mojave Solar	CA	20.0	204.4		Unknown		Proposed	Developer

Name	State	MW - DC	Total area (acres)	Direct area (acres)	Tracking	Module efficiency	Status as of August 2012	Data source
Tuusso Energy Antelope Plant	CA	20.0	211.4		Unknown		Proposed	Third party
Grundman/Wilkinson Solar Farm	CA	21.1*	163.5	117.5	Fixed	11%	Complete	Official documents
Adobe Solar	CA	23.5*	187.9		Unknown		Proposed	Developer
Orion Solar	CA	23.5*	311.2		Unknown		Proposed	Developer
Atwell Island Solar Project	CA	23.5	188.0		Unknown		Construction	Third party
FSE Blythe	CA	25.2	223.2	161.3	Fixed	10%	Complete	Developer
Imperial Valley Solar Company	CA	28.7	153.5		Unknown	15%	Proposed	Third party
McHenry Solar Farm	CA	29.4*	180.9		1 axis	19%	Construction	Developer
Del Sur Solar Project	CA	38.0	219.6		Unknown		Construction	Third party
Lucerne Valley Solar	CA	40.5	495.6	495.6	Fixed	10%	Construction	Official documents
Chocolate Mountains PV Plant	CA	49.9	375.8		Unknown		Construction	Developer
Calipatria Solar Farm 2	CA	50.0	352.4		Unknown		Proposed	Third party
Salton Sea 1	CA	50.0	375.8		Unknown		Proposed	Developer
Avenal SunCity SandDrag Avenal Park	CA	57.7	641.3	442.5	Fixed	9%	Complete	Developer
Copper Mountain PV Plant	CA	58.0	459.2	393.9	Fixed	10%	Proposed	Third party
Midway Solar Farm 1	CA	58.7*	352.4	325.3	Unknown		Proposed	Developer
Regulus Solar	CA	75.0	872.7		Unknown		Proposed	Developer
Calipatria Solar Farm 1	CA	82.2*	352.4	288.9	Unknown		Proposed	Developer
Salton Sea 2	CA	100.0	730.6		Unknown		Proposed	Third party
Quinto Plant	CA	110.0	1,191.0		1 axis	20%	Proposed	Official documents
Imperial Solar Energy Center South	CA	130.0	1,111.1		1 axis	11%	Proposed	Developer
Imperial Solar Energy Center West	CA	150.0	1,241.5		2 axis CPV	25%	Proposed	Developer
Midway Solar Farm 2	CA	182.1*	1,097.5		Unknown		Proposed	Third party
Calexico Solar Farm 1	CA	234.9*	1,468.2		Unknown		Proposed	Developer
Calexico Solar Farm 2	CA	234.9*	1,468.2		Unknown		Proposed	Developer

Name	State	MW - DC	Total area (acres)	Direct area (acres)	Tracking	Module efficiency	Status as of August 2012	Data source
Mount Signal PV Plant	CA	234.9*	1,644.3		Unknown		Proposed	Developer
AV Solar Ranch One	CA	284.0	2,593.0	2,414.0	1 axis	11%	Proposed	Developer
California Valley Solar Ranch	CA	293.6*	2,037.8		1 axis		Construction	Developer
Centinela Solar	CA	323.0*	2,427.7		Unknown		Proposed	Developer
Superstition Solar 1	CA	500.0	6,562.1		Unknown		Proposed	Official documents
Edwards Air Force Base	CA	500.0	3,736.4		Unknown		Proposed	Developer
Desert Sunlight	CA	646.0*	4,985.9	3,529.4	Fixed	10%	Proposed	Official documents
Topaz Solar Farm	CA	646.0*	4,110.8		Fixed	11%	Construction	Developer
Alamosa Water Treatment Facility PV Plant	CO	0.6	6.5	5.6	1 axis	16%	Complete	Official documents
Rifle Pump Station	CO	0.6	5.3	4.3	1 axis	13%	Complete	Official documents
SunEdison Alamosa PV Plant (Fixed-Tilt)	CO	0.6	7.0	3.6	Fixed	14%	Complete	Official documents
Arvada Ralston Water Treatment Plant	CO	0.6	7.1	4.5	1 axis	16%	Complete	Official documents
NREL Mesa Top PV Project	CO	0.7	5.9	3.3	1 axis	16%	Complete	Official documents
SunEdison Alamosa PV Plant (2 Axis)	CO	1.0	14.0	7.3	2 axis flat	14%	Complete	Official documents
NREL National Wind Technology Center	CO	1.1	11.5	7.1	1 axis	13%	Complete	Official documents
Buckley Air Force Base	CO	1.1	4.5	3.8	Fixed	14%	Complete	Official documents
Denver Federal Center Solar Park Phase 1	CO	1.2	7.6	6.0	Fixed	13%	Complete	Official documents
Colorado State University Pueblo Plant	CO	1.2	5.1	4.1	Fixed		Complete	Third party
Denver International Airport Phase 2 (Fuel Farm)	CO	1.6	10.6	8.3	Fixed		Complete	Developer
Rifle Waste Water Reclamation Facility	CO	1.7	14.0	9.9	1 axis	14%	Complete	Official documents
Colorado State University Ft. Collins Phase 1	CO	2.0	17.6	15.0	1 axis		Complete	Third party

Name	State	MW - DC	Total area (acres)	Direct area (acres)	Tracking	Module efficiency	Status as of August 2012	Data source
Denver International Airport 1 Pena Blvd	CO	2.0	11.7	11.7	1 axis		Complete	Developer
Ft. Carson PV Plant	CO	2.0	14.7	12.6	Fixed	11%	Complete	Developer
Colorado State University Ft. Collins Phase 2	CO	3.3	15.4	14.0	Fixed		Complete	Third party
Denver International Airport Phase 3	CO	4.4	35.2	26.9	Fixed		Complete	Developer
Air Force Academy CO Springs	CO	6.0	50.5	31.4	1 axis		Complete	Third party
SunEdison Alamosa PV Plant (1 Axis)	CO	6.6	74.1	38.5	1 axis	14%	Complete	Official documents
Greater Sand Hill Solar Plant	CO	20.0	206.6	132.6	1 axis	20%	Complete	Third party
San Luis Valley Solar Ranch	CO	35.2	258.1		1 axis	20%	Complete	Developer
Cogentrix Alamosa Solar Generating Project	CO	37.0	271.0	224.0	2 axis CPV	31%	Construction	Developer
Kent County Waste Water	DE	1.2	7.0	6.6	Fixed		Construction	Third party
Dover Sun Park	DE	11.7*	121.0	59.1	1 axis	20%	Complete	Third party
NASA PV	FL	1.0	6.1	2.8	Fixed		Complete	Developer
Stanton Energy Center	FL	5.9	41.1	29.1	1 axis		Complete	Developer
Rinehart Solar Farm	FL	8.0	28.2		Unknown	16%	Construction	Third party
Space Coast	FL	11.7*	52.9	35.2	Fixed		Complete	Developer
Jacksonville Solar	FL	15.0	114.4	83.9	Fixed	11%	Construction	Third party
DeSoto Plant	FL	28.0	263.2	201.6	1 axis	19%	Complete	Developer
Sorrento Eagle Dunes phase 1	FL	40.0	164.4		Fixed	14%	Construction	Developer
Sorrento Eagle Dunes phase 2	FL	60.0	422.8		Fixed	16%	Proposed	Third party
Babcock Ranch Solar	FL	75.0	469.8		Unknown		Proposed	Developer
Liberty County Solar Farm	FL	100.0	1174.5		Unknown		Proposed	Third party
Hardee County Solar Farm	FL	200.0	2,349.1		Unknown		Proposed	Third party
Gadsden Solar Farm	FL	400.0	4,698.1		Unknown		Proposed	Third party
Blairsville Plant	GA	1.0	5.7		Fixed		Complete	Third party

Name	State	MW - DC	Total area (acres)	Direct area (acres)	Tracking	Module efficiency	Status as of August 2012	Data source
Kopolei Sustainable Energy Park	HI	1.2	4.7	3.2	Fixed	14%	Complete	Third party
Kalaeloa Oahu	HI	5.0	47.0		1 axis	19%	Construction	Official documents
Exelon City Solar	IL	10.0	48.2	37.6	1 axis		Complete	Developer
Grand Ridge Solar Plant	IL	23.0	187.9		Unknown	12%	Construction	Third party
Indianapolis Airport Solar Farm	IN	10.0	70.5		Fixed		Construction	Third party
Bowling Greens Solar Farm	KY	2.0	15.3	10.6	1 axis		Complete	Third party
William Stanley Business Park	MA	1.9	10.9	7.3	Fixed	14%	Complete	Official documents
Berkshire School	MA	2.0	10.8	9.4	Fixed	15%	Complete	Third party
Northfield Mountain	MA	2.0	12.9	9.3	Fixed		Complete	Third party
Indian Orchard Solar	MA	2.3	14.1		Unknown		Complete	Third party
Springfield Plant	MA	4.2	72.8	47.0	Unknown		Complete	Third party
Mueller Road Holyoke Plant	MA	4.5	22.3		Fixed	15%	Complete	Third party
Canton Landfill	MA	5.7	17.2	12.8	Fixed	15%	Complete	Official documents
Mount St. Mary's University	MD	17.4	158.6	105.7	Fixed	11%	Construction	Third party
Progress Energy	NC	1.2	11.3	9.1	1 axis	14%	Complete	Official documents
Mayberry/Mt. Airy Solar Farm	NC	1.2	7.0		Fixed	14%	Complete	Third party
Neuse River Waste Water	NC	1.3	11.7		Fixed	14%	Complete	Third party
SAS Solar Farm 1 and 2	NC	2.2	20.0	14.1	1 axis	15%	Complete	Developer
Kings Mountain Solar Farm	NC	5.0	32.9		Unknown		Complete	Third party
Murfreesboro	NC	6.4	36.7	30.6	1 axis	19%	Complete	Developer
Davidson County Solar	NC	17.2	221.9	129.3	1 axis		Complete	Developer
Trenton Solar Farm	NJ	1.3	6.5	5.3	Fixed		Complete	Third party
Silver Lake Solar Farm	NJ	2.1	9.4	6.7	Fixed	14%	Complete	Third party
Mars Solar Garden	NJ	2.2	14.4	11.9	Fixed	10%	Complete	Developer
NJMC landfill	NJ	3.0	15.3		Fixed		Complete	Third party
Linden Solar Farm	NJ	3.2	11.7		Unknown		Complete	Third party

Name	State	MW - DC	Total area (acres)	Direct area (acres)	Tracking	Module efficiency	Status as of August 2012	Data source
Janssen Pharmaceutical	NJ	4.1	29.4	21.9	1 axis		Complete	Third party
Vineland	NJ	4.1	32.9	17.6	Fixed		Complete	Developer
Yardville Solar Farm	NJ	4.4	18.4	16.6	Fixed	14%	Complete	Third party
Homdel Solar Farm	NJ	4.8	39.9	18.8	1 axis		Proposed	Third party
Princeton University	NJ	5.3	31.7		1 axis	19%	Complete	Third party
Lawrenceville School	NJ	6.1	35.2		1 axis	15%	Complete	Third party
NJ Oak Solar Farm	NJ	12.5	122.5	97.5	Fixed	14%	Complete	Third party
Upper Pittsgrove	NJ	14.4	105.7		1 axis		Proposed	Third party
Tinton Falls	NJ	19.9	111.6		Unknown		Construction	Third party
Pilesgrove Project	NJ	20.0	148.9	85.3	Fixed	14%	Complete	Third party
Santa Fe Waste Water Plant	NM	1.1	10.4	7.9	1 axis	14%	Complete	Developer
City of Madera Waste Water	NM	1.2	11.2	10.6	2 axis flat	14%	Complete	Third party
Questa	NM	1.2*	20.0	12.6	2 axis CPV	25%	Complete	Third party
Albuquerque Solar Center	NM	2.0	21.7	12.8	Fixed	11%	Complete	Third party
Deming Solar Energy Center	NM	5.0	58.7	40.0	Fixed	11%	Complete	Third party
Alamogordo Solar Center	NM	5.0	58.7		Fixed	11%	Complete	Third party
Hatch Solar Center	NM	6.5	50.1	38.9	2 axis CPV	29%	Complete	Developer
SunEdison Jal	NM	10.7	117.5	86.4	1 axis		Complete	Third party
SunEdison Carlsbad	NM	10.8	100.7	90.3	1 axis		Complete	Third party
Elephant Butte	NM	22.0	187.9		Fixed		Construction	Third party
Roadrunner Solar Facility	NM	23.5*	246.7	198.8	1 axis	11%	Complete	Developer
Cimarron	NM	35.2*	293.6	260.7	Fixed	10%	Complete	Developer
Estancia Solar Farm	NM	50.0	187.9		Unknown		Proposed	Third party
Guadalupe Solar	NM	300.0	2,936.3		Unknown		Proposed	Third party
Las Vegas Solar Center	NV	5.0	58.7		Unknown	11%	Complete	Third party

Name	State	MW - DC	Total area (acres)	Direct area (acres)	Tracking	Module efficiency	Status as of August 2012	Data source
El Dorado Solar	NV	12.0	96.0	84.0	Fixed	11%	Complete	Developer
Nellis Air Force Base	NV	18.0	186.7	148.0	1 axis		Complete	Official documents
Searchlight Solar Project	NV	20.0	242.3		1 axis		Complete	Third party
Fish Springs	NV	20.6	211.4		Fixed	10%	Construction	Official documents
Apex	NV	24.9	187.1		Unknown		Proposed	Third party
Silver State Solar North	NV	65.1	775.0		Fixed	10%	Construction	Official documents
Boulder City	NV	176.2*	1,303.7		1 axis	10%	Construction	Developer
Silver State Solar South	NV	350.0	3,406.1	3484.8	1 axis	10%	Proposed	Official documents
Mojave Green Center	NV	720.0	6,384.7		Unknown		Proposed	Third party
Brookhaven Lab	NY	37.0	231.3	225.5	Fixed	13%	Construction	Developer
Washington Township Solar Array	OH	1.1	11.5	8.4	Fixed	9%	Complete	Developer
BNB Napoleon Solar LLC	OH	9.8	70.5		1 axis	19%	Construction	Third party
Wyandot Solar	OH	12.6	97.0	78.0	Fixed	11%	Complete	Developer
Turning Point Solar	OH	58.6*	496.4		Unknown		Proposed	Official document
Yamhill Solar	OR	1.2	11.0		Fixed	10%	Complete	Developer
Bellevue Solar	OR	1.7	14.0		Fixed	10%	Complete	Developer
Pocono Raceway	PA	3.0	27.2	17.9	Fixed		Construction	Third party
Exelon Conergy	PA	3.0	19.4	12.9	Fixed		Complete	Developer
Claysville Solar Project	PA	20.0	117.5	99.5	Fixed		Proposed	Developer
Shelby Solar Project	SC	1.0	10.6	6.5	1 axis	19%	Complete	Third party
West Tennessee Solar Farm	TN	5.0	29.4	26.9	Fixed		Construction	Developer
Blue Wing Solar	TX	16.1	124.2	95.7	Fixed		Construction	Developer
Austin Energy Webberville	TX	34.3	434.3		Unknown	15%	Complete	Third party
Pflugerville Solar	TX	60.0	704.7		Unknown		Construction	Third party
South Burlington Solar Farm	VT	2.2	31.7	25.8	2 axis flat		Complete	Third party

Appendix C. CSP Projects Evaluated

Table C-1. Concentrating Solar Power Land-Use Data

Note: Additional CSP plant information, such as storage and annual generation, can be found in Appendix A.

Name	State	MW - AC	Total area (acres)	Direct area (acres)	Technology	Status as of August 2012	Data source
Maricopa Solar Project	AZ	1.5	15	4	Stirling Engine	Complete	Third party
Quartzsite	AZ	100	1,675		Tower	Proposed	Developer
Crossroad Solar	AZ	150	2,560		Tower	Proposed	Developer
Solana	AZ	280	1,920		Parabolic trough	Construction	Third party
Sierra SunTower	CA	5	50	22	Tower	Complete	Developer
Kimberlina	CA	7.5	35	15	Linear Fresnel	Complete	Developer
Solar Two	CA	10	132	110	Tower	Decommissioned	Third party
Coalinga	CA	13	86	57	Tower	Proposed	Developer
Victorville 2 hybrid	CA	50	265	230	Parabolic trough	Proposed	Official document
Palmdale Hybrid Gas/solar Plant	CA	57	377	250	Parabolic trough	Proposed	Official document
Rice Solar	CA	150	2,560	1,410	Tower	Construction	Official document
Abengoa Mojave	CA	250	1,765		Parabolic trough	Construction	Third party
Ford Dry Lake (Genesis)	CA	250	4,640	1,800	Parabolic trough	Construction	Official document
Hidden Hills 1	CA	250	1,640	1,560	Tower	Proposed	Official document
Hidden Hills 2	CA	250	1,640	1,560	Tower	Proposed	Official document
Rio Mesa 1	CA	250	1,917		Tower	Proposed	Official document
Rio Mesa 2	CA	250	1,917		Tower	Proposed	Official document
Rio Mesa 3	CA	250	1,917		Tower	Proposed	Official document
SEGS (all)	CA	354	2,057	2,057	Parabolic trough	Complete	Third party
Ivanpah All	CA	370	3,515	3,236	Tower	Construction	Official document
Saguache Solar	CO	200	3,000	2,669	Tower	Construction	Official document
Martin Next Generation	FL	75	500	400	Parabolic trough	Complete	Developer

Name	State	MW - AC	Total area (acres)	Direct area (acres)	Technology	Status as of August 2012	Data source
Nevada Solar One	NV	64	400	290	Parabolic trough	Complete	Third party
Tonopah (Crescent Dunes)	NV	110	1,600	1,527	Tower	Construction	Developer

33

Appendix D. Impact of PV System Size and Module Efficiency on Land-Use Requirements

System size appears to have little impact on capacity-based land-use requirements. Figure D-1 and Figure D-2 show the total-area requirements for small and large PV systems, with respect to project capacity. No significant trends are observed for land use and system size for small or large PV systems.

Land use was also evaluated with respect to module efficiency. Figure D-3 shows capacity-based direct land-use requirements for all PV systems with respect to module efficiency, and Figure D-4 shows the generation-based direct land-use requirements. We expect that land use will decrease with increasing module efficiencies, but no significant trends are observed for land use and module efficiency for small or large PV systems. A linear regression analysis yields a poor correlation coefficient for both the capacity-based area data (0.04) and the generation-based data (0.08). Isolating for fixed-tilt systems reveals that projects with higher efficiency use less land on a capacity basis (with a correlation coefficient of 0.50). No trends are observed within the pool of 1-axis tracking systems. Variations in land use that remain after isolating for module efficiency and tracking type are not clearly understood.

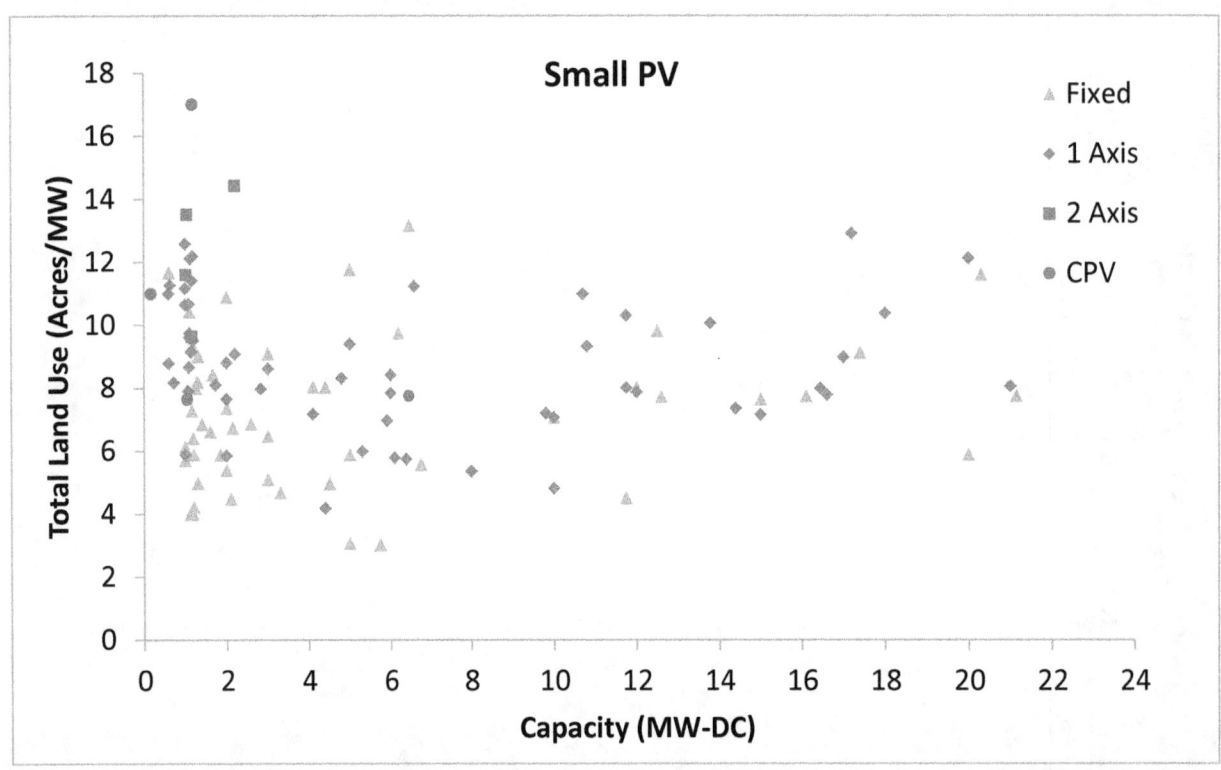

Figure D-1. Total-area requirements for small PV installations as a function of PV plant size

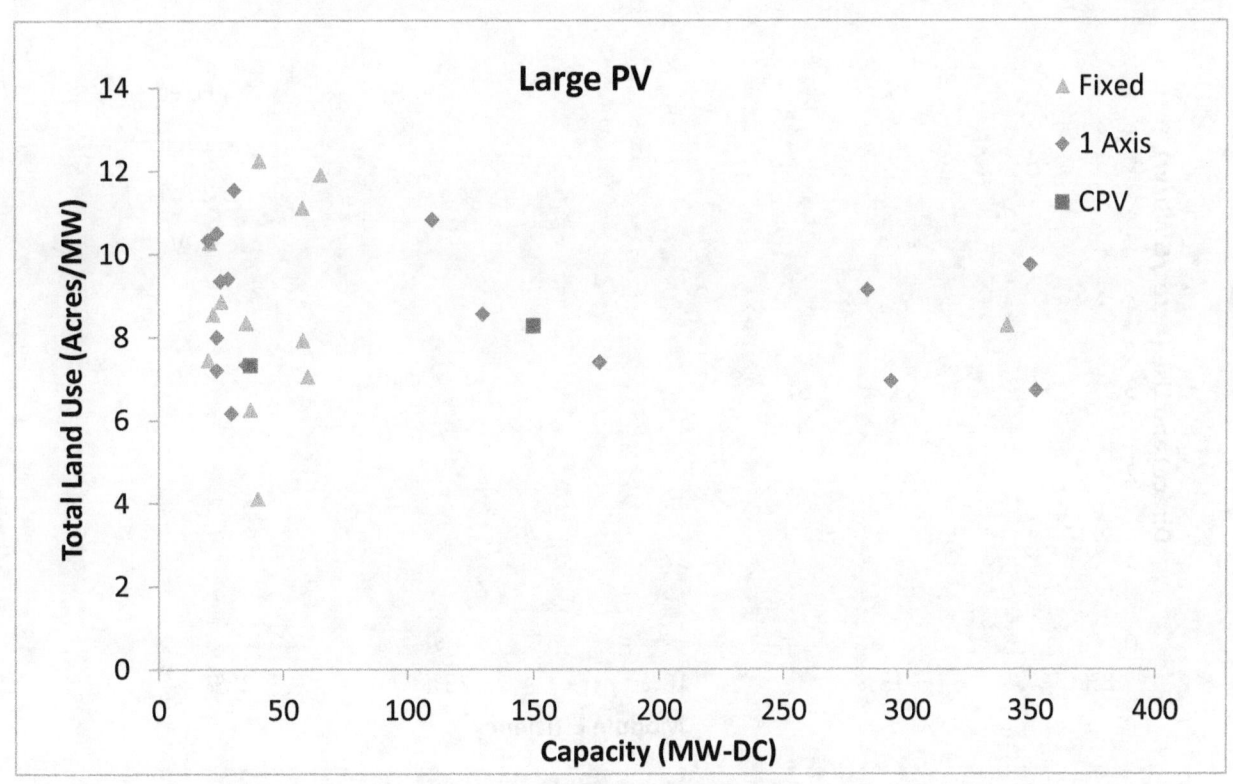

Figure D-2. Total-area requirements for large PV installations as a function of PV plant size

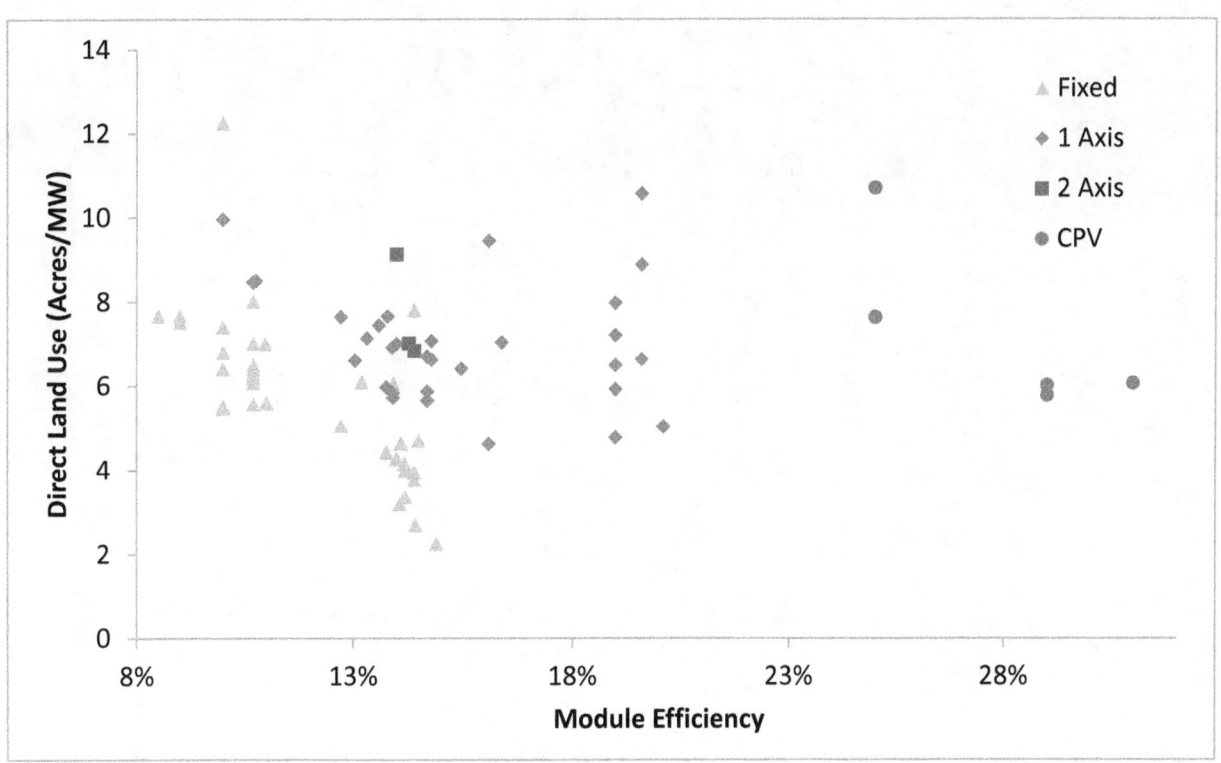

Figure D-3. Capacity-based direct-area land-use requirements for all PV systems as a function of module efficiency

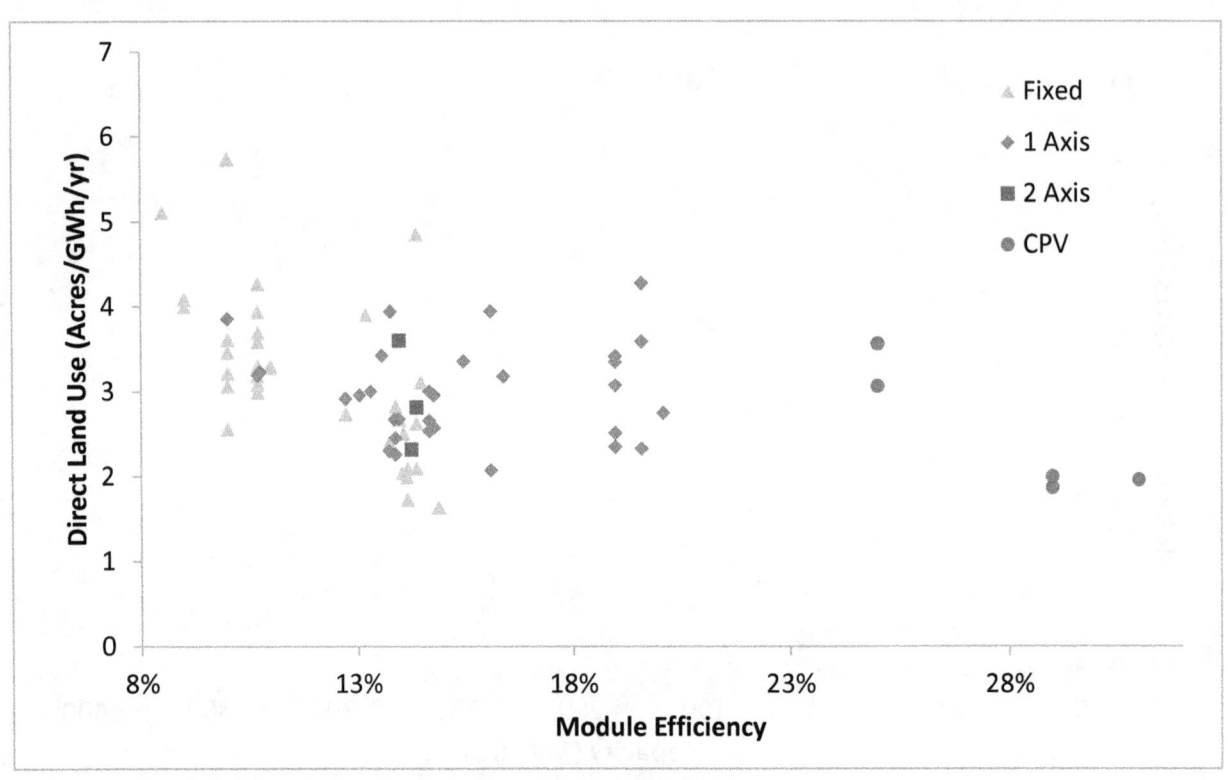

Figure D-4. Generation-based direct-area land-use requirements for all PV systems as a function of module efficiency

Appendix E. Impact of CSP System Size and Storage on Land-Use Requirements

We evaluated the impact of project capacity on land-use requirements and found that system size appears to have little impact on generation-based CSP land-use requirements. Figure E-1 and Figure E-2 show the total-area and direct-area requirements for all CSP systems evaluated, with respect to system size. No significant trends are observed for land-use and capacity for CSP systems.

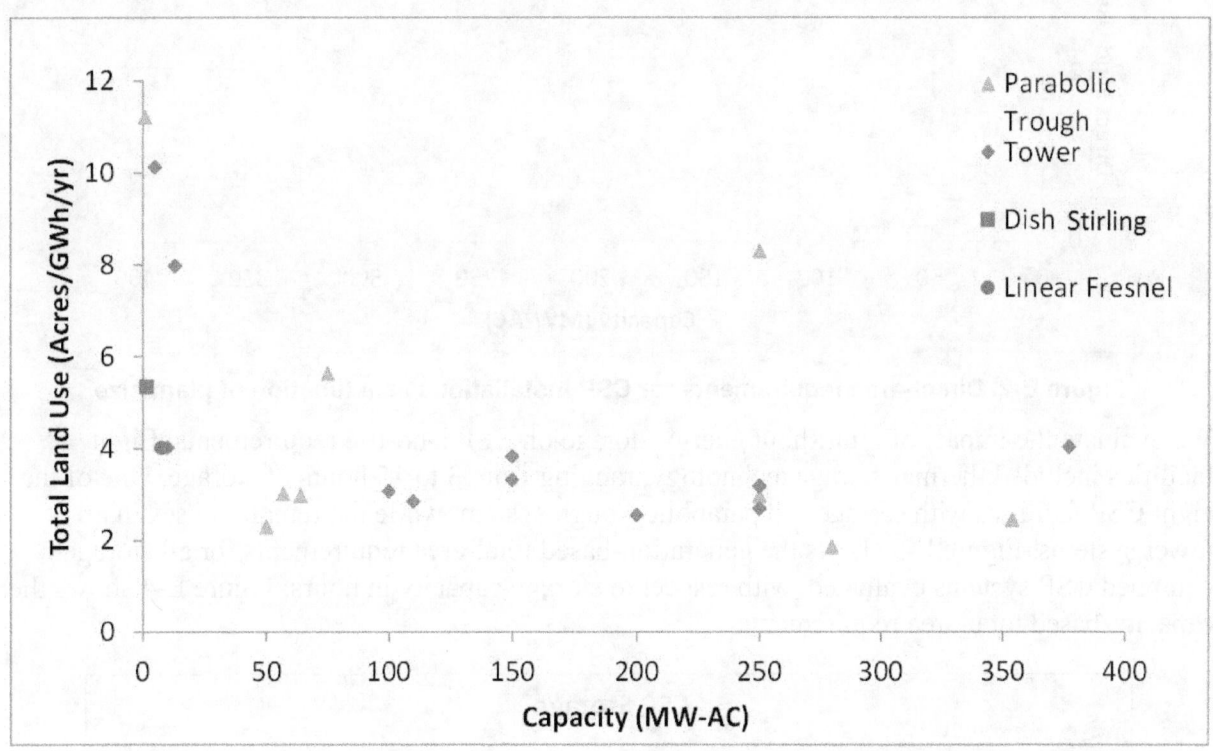

Figure E-1. Total-area requirements for CSP installations as a function of plant size

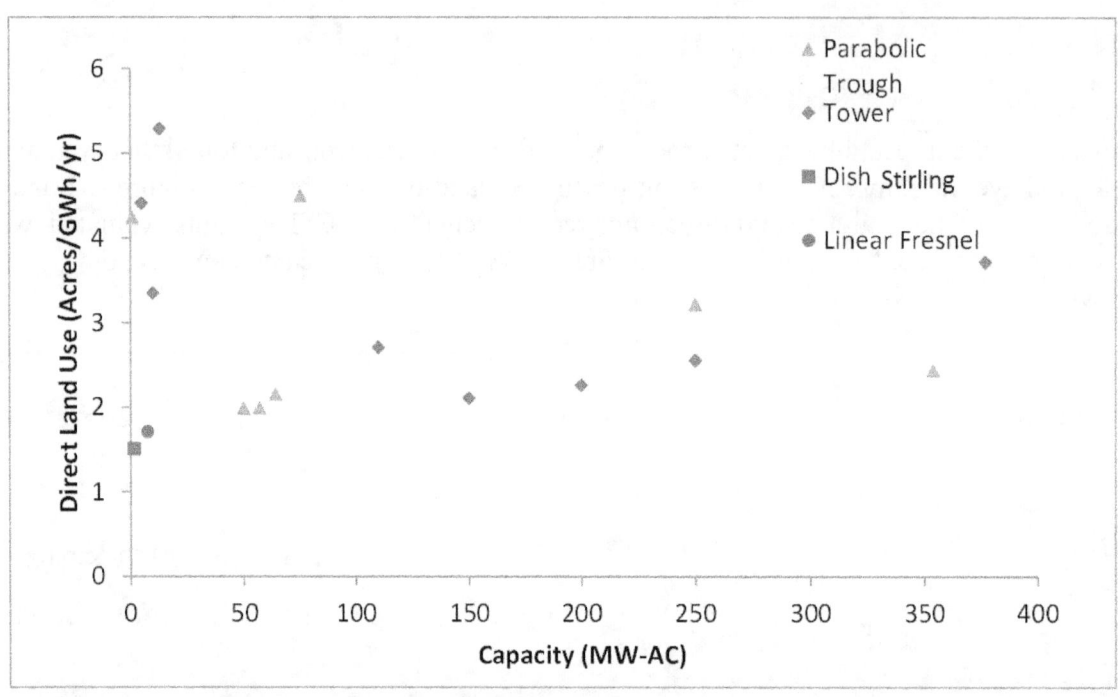

Figure E-2. Direct-area requirements for CSP installations as a function of plant size

We evaluate the impact of multi-hour energy storage on CSP land-use requirements. Eight facilities included thermal storage technology, ranging from 3 to 15 hours of storage. One of the eight CSP facilities with storage is a parabolic trough system, while the remaining seven are tower systems. Figure E-3 shows the generation-based total-area requirements for all storage-equipped CSP systems evaluated, with respect to storage capacity in hours. Figure E-4 shows the capacity-based total-area requirements.

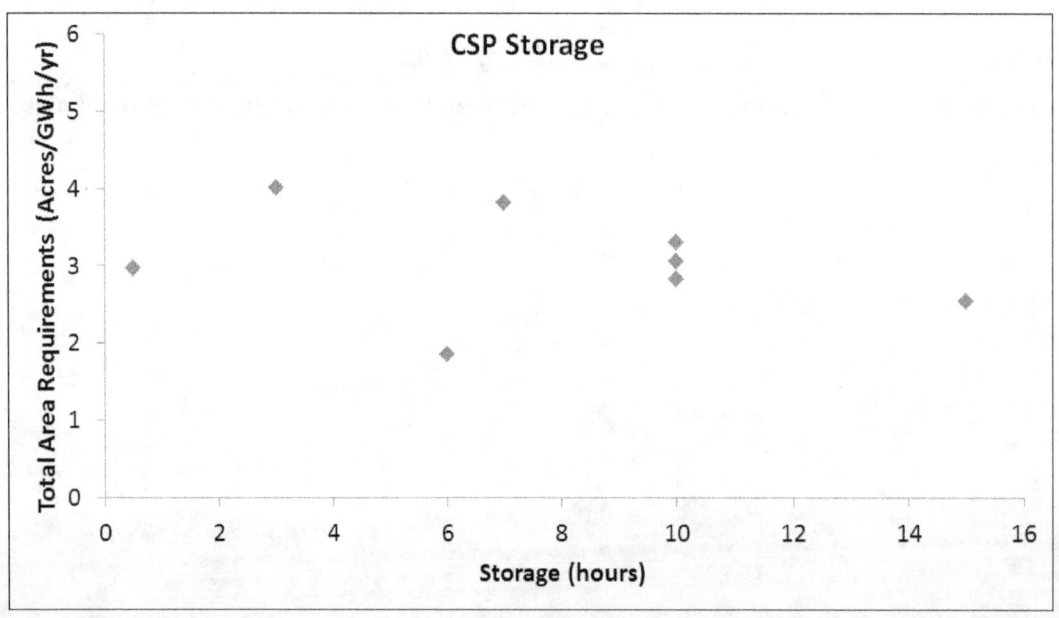

Figure E-3. Total generation-based area requirements for CSP installations as a function of storage hours

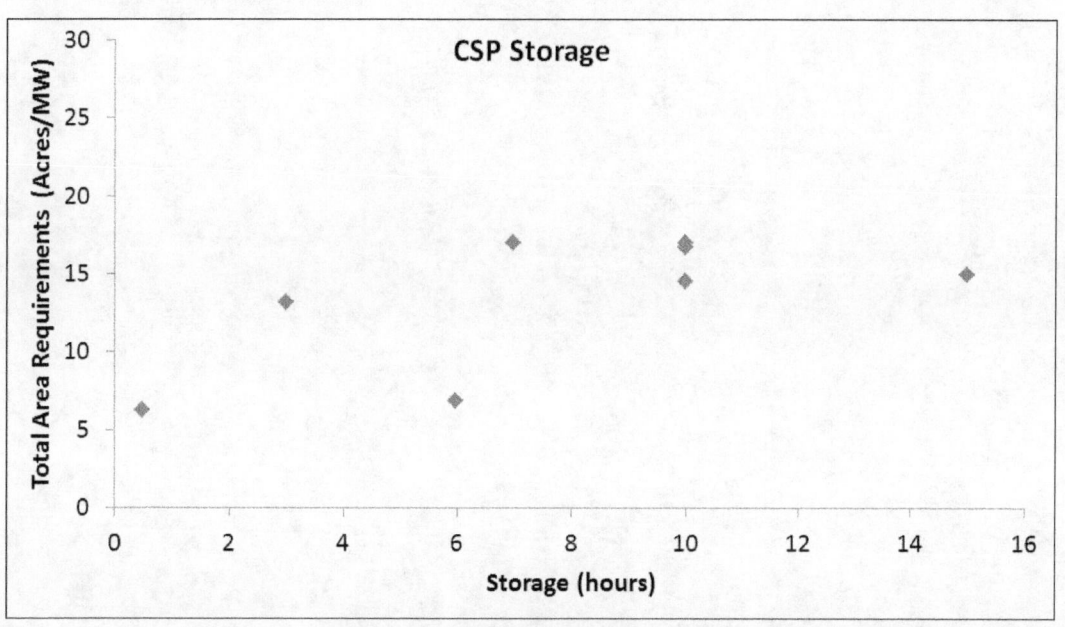

Figure E-4. Total capacity-based area requirements for CSP installations as a function of storage hours